THE DANCING BEES

# THE

# DANCING

# BEES

## KARL VON FRISCH

### AND THE DISCOVERY OF THE

### HONEYBEE LANGUAGE

## TANIA MUNZ

*The University of Chicago Press*

CHICAGO AND LONDON

The University of Chicago Press, Chicago 60637
The University of Chicago Press, Ltd., London
© 2016 by The University of Chicago
All rights reserved. No part of this book may be used or reproduced in any
manner whatsoever without written permission, except in the case of brief
quotations in critical articles and reviews. For more information, contact the
University of Chicago Press, 1427 E. 60th St., Chicago, IL 60637.
Published 2016
Paperback edition 2017
Printed in the United States of America

23  22  21  20  19  18  17      4  5  6  7  8

ISBN-13: 978-0-226-02086-0 (cloth)
ISBN-13: 978-0-226-52650-8 (paper)
ISBN-13: 978-0-226-02105-8 (e-book)
DOI: https://doi.org/10.7208/chicago/9780226021058.001.0001

Frontispiece: see fig. 8.2, p. 219, for more information.

Library of Congress Cataloging-in-Publication Data

Names: Munz, Tania, 1973– author.
Title: The dancing bees : Karl von Frisch and the discovery of the honeybee
     language / Tania Munz.
Description: Chicago ; London : The University of Chicago Press,
     2016. | Includes bibliographical references and index.
Identifiers: LCCN 2015036905 | ISBN 9780226020860 (cloth : alk.
     paper) | ISBN 9780226021058 (e-book)
Subjects: LCSH: Frisch, Karl von, 1886–1982. | Bees—Behavior.
Classification: LLC QL31.F7 M959 2016 | DDC 595.79/9—dc23 LC record
     available at http://lccn.loc.gov/2015036905

♾ This paper meets the requirements of ANSI/NISO Z39.48-1992
(Permanence of Paper).

*To*
*my parents.*

*And for*
*Tim.*

The life of the bees is a magic well.
The more one draws from it, the more richly it flows.

KARL VON FRISCH,
*Aus dem Leben der Bienen* (1953)

# Contents

# Sensational Findings

In January of 1946, while much of Europe lay buried under the rubble of World War II, Karl von Frisch penned an excited letter to his friend and fellow animal behaviorist Otto Koehler. In the letter, the Austrian-born experimental physiologist and bee researcher reported his "sensational findings about the language of bees." Over the previous two summers, he had discovered that honeybees communicate to their hive mates the distance and direction of food sources by means of the "dances" they perform after returning from foraging flights. He found that the insects indicate nearby food sources via a circular dance and faraway foods via a figure-eight-shaped waggle dance. The straight-run portion of the waggle dance, he explained, also contains information about direction, and the frequency of its turns correlates closely with distance: the closer the supply, the more rapidly the bees dance. The otherwise reserved von Frisch concluded his missive, "And if you now think I'm crazy, you'd be wrong. But I could certainly understand it."[1]

Von Frisch was right—his discovery was a sensation, and news of it quickly spread throughout Europe and abroad. Although his interpretation of bee communication would at times face fierce opposition, the work endured as a classic example of animal behavior and succeeded in placing bees alongside primates, dolphins, and birds in the mid-twentieth-century pantheon of communicating beasts. The postwar period saw a flurry of activity: interdisciplinary study groups, edited volumes, and conferences all aimed to tackle the problem of animal communication.[2] Honeybees held a prominent place in this research. During the 1960s, their "language" was the most widely studied form of animal communication, and some deemed it the most complex, second only to human speech.[3] The stakes were especially high, as language had long been considered a window into human minds and souls, and one of the key distinctions between humans and animals.

Von Frisch's findings and approach were praised by many of the period's best-known scientists, including the sociobiologist and ant specialist E. O. Wilson, the cognitive ethologist Donald Griffin, and the ethologist and Nobel laureate Nikolaas Tinbergen. Generations of bee researchers would go on to discuss and cite his work. And in 1973, he was awarded one of the highest honors bestowed on scientists—his findings earned him a share of the Nobel Prize for Physiology or Medicine, together with Tinbergen and fellow Austrian Konrad Lorenz.[4]

Even before his receipt of the Nobel Prize, von Frisch's reputation had already reached far and wide. Throughout much of his career, he wrote books and articles for lay readers and edited a series of small books under the title "Understandable Science" that specifically aimed to bring the work of well-known scientists to the German-speaking public. He gave talks to audiences that ranged from school-age children to retirees and produced a number of pioneering films that used ingenious methods to show the sensory world of animals.[5] With his boyish build and thick wire-rimmed spectacles, he introduced young and old to the mysterious world of bees.

*The Dancing Bees* is a dual biography—on the one hand, of von Frisch as one of the most innovative and successful scientists of the twentieth century and, on the other hand, of his honeybees as experimental and, especially, communicating animals that play a rich role in human culture. It tells of von Frisch's discovery of the dance language in the context of the politics and events that changed Germany and the world over the course of the twentieth century. Others before him, going as far back as Aristotle, had observed the animals' dances, and beekeepers had long speculated about the possibility of some form of bee communication to mobilize foragers to exploit food sources. But von Frisch was the first to definitively link the bees' dances to their recruitment abilities.

How, we wonder, did von Frisch manage to observe such behaviors where others for centuries had looked but failed to see?

Alongside von Frisch's story, I include a series of vignettes that explore aspects of the bees' lives that illustrate the transition the animals underwent in the popular and scientific imaginations over the course of the century. Long before von Frisch revealed the insects' remarkable dance language, humans and bees had shared an entwined and complex history. From Virgil's poetry of the first century BC to Bernard Mandeville's eighteenth-century, subtly subversive fable of the insects' politics, the hive served as inspiration for how a well-run social polity might function and proved a ready canvas for moral projections onto nature.⁶ I write the bees' history alongside human history to reveal the changing cultural and scientific understandings of the insects as they transitioned from political to physiological and, finally, to communicating animals.

And yet while students across the world continue to learn about the honeybee dance language, less well known today are the circumstances under which von Frisch's discoveries were made. As the date of the letter reporting the "sensational findings"—January 1946—suggests, he performed some of his most important work against the backdrop of the deadliest conflict in human history.

When Hitler took power in 1933, von Frisch was professor of zoology at the University of Munich. Soon after, the city and its institutions were brought under Nazi control through the process of *Gleichschaltung* (coordination).⁷ In an effort to purge government institutions of Jews and other "undesirables," the Nazi government required all civil servants, including professors, to furnish proof of their Aryan descent. In the case of von Frisch, what had begun as vague rumors about him crystallized by late 1940 into ominous fact: after months of searching, the Nazi office responsible for genealogical research located evidence that his maternal grandmother had been of Jewish descent. Like many nineteenth-century assimilated Jews who considered themselves loyal citizens of the *Kaiserreich*, her parents had converted to Catholicism just a few years before she was born.⁸ Presumably the young couple had done so in the hope of securing a better future for their famly in a Christian-dominated society. But the Nazi concept of what it meant to be Jewish was overwhelmingly rooted in notions of blood and descent rather than cultural heritage or religious belief. As a consequence, people like von Frisch, who had never before considered themselves Jewish, were now officially declared such based on the more or less distant tendrils of their ancestry.

The fate of the "Quarter Jews" was never as dire as that of the "Half" and "Full Jews."[9] But gradually they were excluded from higher education and certain professions, such as medicine and law. In the case of von Frisch, the declaration of his status as Quarter Jew by the president of the University of Munich in early 1941 also announced that he was to be ousted from his position at the university. Von Frisch was devastated. He was only fifty-four years old and lived for his work.

Prior to the war and indeed throughout most of his life, he had considered himself staunchly apolitical. Nonetheless, his instincts about whom to turn to and how best to position his work in the scientific landscape of the times would serve him well. Through an odd twist of fate, he performed some of his most important work not so much despite the regime as because of it. And the honeybees played a crucial role.

At the very moment that von Frisch received news of his impending ouster, his favorite research subjects—the honeybees— were also in trouble. A small intestinal parasite, known by its Latin name *Nosema apis* (or *Nosema*, for short), was wreaking havoc on the insects' insides. The invaders multiplied so prolifically that they quite literally burst the animals' guts. Afflicted bees crawled to their deaths leaving behind a trail of feces. During 1940 and 1941, hundreds of thousands of hives across Germany were destroyed by the parasite.[10] Although the causes of the sudden outbreak were unknown, it was clear that the loss of agriculture's main pollinators would spell disaster for humans, and especially for a weary German nation at war.

By winter of 1941, the stunning results of the blitzkrieg were giving way to a grinding war of attrition. Now, it seemed, the war could not be won by brief displays of force but by the nation that could keep its men armed and fed the longest. Germany redoubled its efforts. From military strategy to the economy and from agricultural policy to propaganda, all resources bled into the war machine. And as the German nation bore down for total war, von Frisch would gain a small foothold in the Nazi bureaucracy. The hungry nation at war depended critically on agriculture to feed bodies in battle and mouths on the home front.[11] Agriculture depended on the successful cultivation and growth of food crops, and many of these crops relied on honeybees; only blooms pollinated by the insects would bear fruit. Von Frisch was classically trained in experimental physiology and knew precious little about intestinal parasites and even less about large-scale agriculture. But he knew an awful lot about bees.

When I first read von Frisch's work, I was drawn in by the elegant simplicity and compelling logic of his experiments. He seemed able to distill the potentially overwhelming complexity of the outdoors into something that more closely resembled the clean and orderly world of the laboratory. And yet the more I read in his papers, letters, and books, the more I found myself wanting to know about the messy world that he seemed to keep at bay when working with his bees. As I followed him on his path, a far less tidy picture emerged; his work, it turned out, was very much of its time and place. Shortly after he received news of his impending ouster, he appealed to friends and colleagues for help. He also switched the focus of his research to a more practical orientation in the hope that the Nazi authorities would deem his work worth continuing. What, I wondered, are we to make of the fact that he performed his most important research while receiving funding from the Nazi Ministry of Food and Agriculture? How should such choices be evaluated in light of what we know today about Nazism and its relationship to science? And what were the postwar consequences of his having been declared a Quarter Jew by the Nazis?

Over the years that I immersed myself in von Frisch's life and work, I learned that his story resists simple answers. Instead, it offers us a rare glimpse into how he navigated the overwhelming and often frightening Nazi bureaucracy. In considering von Frisch's biography, we are forced to reject simplistic accounts that paint a regime hostile to all but the crudest and most racialized sciences. His work was neither racially motivated nor pseudoscientific (like much of Nazi genetics, anthropology, psychology, medicine, and physics). And yet the research fit in well with Germany's long-held aspirations of gaining independence from the import of foods and raw materials to prepare for and wage war.[12]

After the war, von Frisch once again reached out to his colleagues across the Atlantic and Channel. His wartime fate played no small part in easing this contact. Scientists who had worked in the Reich during the war often were excluded or given a chilly reception by their colleagues and old friends in the former Allied countries. But because it became known that the Nazis had declared von Frisch one-quarter Jewish and threatened his job, his political credentials were soon considered beyond reproach; a perceived enemy of the Nazis could easily be considered a friend of the Allies.[13] To Americans who were looking for "good Germans" in the wake of the war, especially as their attentions in the late 1940s shifted from denazification to a cold war with the Soviet Union that was heating up, this political exoneration made him an attractive candidate for support and funding. On a 1949 lecture tour to the United States, von Frisch reestablished important connections with access to the deep pockets of the Rockefeller Foundation. These transatlantic allegiances would also support him during heated debates about the meaning of the bee dances in the 1960s and 1970s.

At issue in the dispute that started in the 1960s was not merely whether the bees' dances served to communicate information about the location of foods but rather the very approach to animals. Were nonhuman animals capable of symbolic communication? And if so, what did this mean for a scientific understanding of the animal-human boundary?

Von Frisch's approach and answers to these questions ultimately withstood the challenge, and the receipt of the Nobel Prize was a testament to a science of animals that had come into its own. Animal behavior studies had successfully shed its nineteenth-century reputation of lacking objectivity and reveling in anecdotes about morally upstanding worker bees, wise queens, and lazy drones. And yet in their place emerged a new kind of model for humanity. When the Rockefeller Foundation once again sponsored von Frisch's work in the postwar period, it did so with the explicit desire to understand communication across peoples and cultures in the wake of the war that went down in history as one of humanity's greatest disasters.

Even today, as science has advanced to give us state-of-the-art access to the hive and its inhabitants, bees still occupy an important cultural and political space. As we must once again face the sobering prospects of the dying bees, the hive serves as a mirror that reflects back to us the values, fears, and aspirations of our time. Colony collapse disorder brings us yet again face-to-face with both the limits of our scientific reach and our dependence on these creatures for our food supply. What is causing the bees to die?[14] We do not know exactly. But we assume that their catastrophically increased death rates are somehow related to our modern ways of life—possible contenders in the search for explanations range from the stresses associated with monocultural factory farming and pesticide exposure, to overuse of antibiotics, to the introduction from Asia of the parasitic *Varroa destructor* mite. Thus, despite our ongoing efforts to purge modern science of the anthropomorphic pull of previous generations, bees continue to challenge our very understandings of the human "us" and animal "them."

# VICTORIAN BEES

In the darkness of the hive, tens of thousands of bees go about their business. Some tend the young while others clean cells. Still others repair and build combs while a few bees guard the entrance from intruders. Another set meets foragers as they return from their nectar-gathering flights. These insects receive their sisters' regurgitated food and move it into storage cells. Other gatherers walk deep into the hive with hind legs laden with pollen and kick off the bounty of their forage into dedicated cells. A handful of bees face the queen as she moves through the dense tangle of bodies and dips her abdomen into cell after cell. Each time, she deposits a single tiny egg—glossy and translucent, like a perfect grain of rice. After three days, the eggs quiver and give way to minute, wormlike larvae. They lie curled in their cells and eat. And eat. Nine days later, the larvae weigh more than a thousand times what the original egg weighed. Worker bees now seal the larvae's cells. The enclosed creatures enter the pupal stage, in which they neither eat nor drink. On the twenty-first day, the transformation is complete and freshly formed worker bees emerge. Such are the workings of the hive and the stuff of bee lore over the centuries.

Bees have long captured the attention of not just beekeepers and naturalists but also economists, poets, political scientists, and social theorists. The animals have been admired for their work ethic, statecraft, and gifts of honey and wax. But above all else, commentators have wondered about the little animals' remarkable abilities to manage their complex social world. If no single animal—not even the queen—can oversee the whole, then how does the hive arrive at decisions that ensure its survival? For naturalists, bees came to stand at the very pinnacle of instincts as animals whose every move was determined by complex inborn programs of behavior.

If naturalists crowned the bees supreme mistresses of instinct, they appreciated the insects' comb building as an especially impressive example in their repertoire of accomplishments. So smitten was Charles Darwin with the animals' elegant combs that he declared him "a dull man who can examine the exquisite structure of a comb, so beautifully adapted to its end, without enthusiastic admiration." Dull, indeed. Stacked row upon row, in perfect, machinelike fashion, the cells were deemed marvels of geometric efficiency. The shapes nested in such a way as to distribute their weight equally between neighboring cells. Each hexagon shares its walls with adjacent cells, thereby requiring the least amount of wax for its construction. The eighteenth-century mathematician Colin Maclaurin wrote of the cells' shape: "What is most beautiful and regular, is also found to be most useful and excellent."[1]

And yet while all could agree that the combs were elegant and excellent, opinions diverged over what might be required of the bees to execute these constructions. To many, the combs' seemingly high degree of sophistication demanded an explanation beyond the merely natural; indeed, they believed them to evidence God's divine plan in nature. The early modern philosopher Thomas Reid argued that the bees "work most geometrically, without any knowledge of geometry; somewhat like a child, who, by turning the handle of an organ, makes good music, without any knowledge of music." He concluded: "The geometry is not in the Bee, but in the great Geometrician who made the Bee, and made all things in number, weight, and measure."[2]

Darwin is widely hailed as having put an end to such appeals to God to explain structure and function in nature. But how, if not by the hand of God, had animals and plants become so exquisitely adapted to their surroundings? According to Darwin, instincts, like physical structures, varied and were heritable and therefore subject to natural selection. Since some instincts served organisms better than others, those animals that possessed them would presumably live longer and leave more offspring than their less fortunate kin, and the traits that had offered them an edge in survival and propagation would be passed on to their sons and daughters. Darwin knew well that the success or failure of his theory of evolution depended critically on its ability to explain such complex behaviors as comb building in bees. And indeed, he dedicated considerable space and energy in his *Origin of Species* to describing just how the comb-building instinct might have arisen from more primitive bees that bored holes into wax.[3]

Yet even among those who cheered Darwin's efforts to naturalize explanations of the living world, some observers asked whether there might not be something more—perhaps closer to judgment—that guided the animals' mysterious behaviors? By the late nineteenth century, at least one naturalist was willing to put a name to this something—intelligence. The English comparative psychologist George John Romanes in his 1882 book, *Animal Intelligence*, surveyed the animal kingdom from mollusk to primate to marshal examples in support of the book's title. In a chapter dedicated to bees and wasps, he told of many wondrous feats that he considered too spectacular to be left to mere instincts. Indeed, he argued that these examples *proved* the existence of the very kinds of foresight and insight that characterize intelligence in humans.

He too considered the honeybees' combs "the most astonishing products of instinct that are presented in the animal kingdom." But, he continued, it was "an instinct not wholly of a blind or mechanical kind but [one that] is constantly under the control of intelligent purpose." In support of this claim, he cited a range of observers who had witnessed the animals making subtle adjustments to their comb building with seemingly humanlike intelligence. He cited one well-known apiculturist who had witnessed some bees tearing down and then correcting sections of comb that their sisters had built, because they found the structures wanting. In another instance, after a piece of comb had fallen to the ground, the bees not only repaired the missing piece but also reinforced other sections of the comb to prevent future calamities. To Romanes, such adjustments evidenced foresight and insight that contradicted the blind execution of instinct that others had postulated.[4]

Examples of a less savory kind also impressed Victorian onlookers as all too "human" and therefore indicative of faculties beyond mere animal instinct. With regret, they recounted lurid examples of bees engaged in warfare, theft, and murder. Not surprisingly, such descriptions of the animals' darker side were often saturated with moral opprobrium. When bees with the "thieving propensity" were observed entering a foreign hive, their intentions could be clearly read. According to Romanes, "They show by their whole behavior—creeping into the hive with careful vigilance—that they are perfectly conscious of their bad conduct; whereas the workers belonging to the hive fly in quickly and openly, and in full consciousness of their right." The behavior soon spreads, and "the whole bee-nation may develop marauding habits, and when they do this they act in concert to rob by force."[5]

If such accounts called into question the bees' moral fiber, descriptions of their emotional makeup offered even less comfort. Sir John Lubbock, one of the nineteenth century's foremost experts on insects, painted a thoroughly grim picture: "Far, indeed, from having been able to discover any evidence of affection among them, they appear to be thoroughly callous and utterly indifferent to one another." On one occasion, he reported having been forced to kill a bee for experimental purposes. He crushed the unfortunate animal right next to one of her hive mates. So close were the animals that their wings touched, and yet "the survivor took no notice whatever of the death of her sister, but went on feeding with every appearance of composure and enjoyment, just as if nothing had happened."[6]

Despite such chilling accounts of the insects' indifference toward their hive mates, the animals nevertheless seemed to display remarkable levels of cooperation. Indeed, it was precisely this ability to settle the complex affairs of the hive communally that prompted commentators to wonder about their ability to communicate. For example, observers offered vivid accounts of the collective reaction to the loss of a queen. Since a queenless hive is as good as a dead hive, the "news" of the loss soon pitches the hive into a frenzy. The German bee researcher Ludwig Büchner reported how "in a little time . . . the sad event will be noticed by a small part of the community, and these will stop working and run hastily about over the comb." Soon, the "excited bees . . . leave the little circle in which they at first revolved, and when they meet their comrades they cross their antennae and lightly touch the others with them. The bees which have received some impression from this touch now become uneasy in their turn, and convey their uneasiness and distress in the same way to the other parts of the dwelling."[7] Thus, according to Büchner, the busy bees disseminate the bad tidings with the swiftness of village gossips.

But not everyone agreed that the bees communicate. The utterly sensible Sir John Lubbock, for example, reported to the British journal *Nature* that his experiments had yielded no evidence of the phenomenon. He had offered food to one of his bees in a location that lay out of sight of the hive. While feeding the animal, he had marked her to see whether she would return and recruit her hive mates to the hidden location. He waited, but to no avail—though the marked bee appeared repeatedly at the food source, her sisters never showed up.[8]

More typical, however, was the view of an American by the name of Josiah Emery who in a letter to *Nature* countered Lubbock's skepticism. He recounted the methods of American "bee hunters" as evidence of the bees' abilities to communicate. A hunter in search of wild honey, he explained, would first capture a single foraging bee and place her in a box with honey. After allowing the bee to "gorge" herself, he would release her and wait until she reappeared with some of her hive mates. The hunter would now catch these animals as well and place them in the box so that they too could feed on the honey. Now he could move around the area and release the insects one by one, each time paying close attention to the direction in which they flew. As the animals returned in "beeline" to their hive, the hunter could determine the location of the bees' home by triangulation. The letter writer urged that when the bee hunters released their first-caught bee and waited, they relied precisely on the bees' ability to communicate with and recruit their hive mates to food stores. "It is possible that our American bees are more intelligent than European bees," he quipped to a largely

British readership, "but hardly probable; and I certainly shall not ask an Englishman to admit it."[9]

Were American bees more intelligent than British bees? Unlikely. But it would be another century until careful research revealed by just how great a margin the sober Sir John Lubbock had underestimated the bees' abilities to communicate.

# Coming of Age in Vienna

Karl's father, Anton von Frisch, was a nineteen-year-old medical student when he first met Marie Exner. She was the sister of a friend who'd invited him to spend part of his summer vacation in the beautiful Salzkammergut region of Austria. For the previous five summers, Marie and her siblings had left Vienna to spend their holidays there, either by invitation or, as they got older, by renting accommodations. The five Exner children had been orphaned almost a decade earlier when the youngest boy was only ten. The siblings closed ranks after the tragedy and formed a close-knit unit. Despite growing up with guardians in separate homes, they remained close throughout their adolescence and adulthood. That summer, in 1868, they rejoiced at once again being united. Spirits were especially high as various friends, including Anton, joined them to swim in the nearby lake and roam the wooded mountains.[1]

Marie, aged twenty-four, was the only girl of the bunch and the middle child. According to family lore, the attraction between her and Anton was as immediate as it was overwhelming. But the two kept their affections secret at first, even from each other. Marie, as Anton soon learned, was already engaged to a young philosopher and family friend, Ludwig Zitkovsky, who was hoping to secure a teaching position so they could marry. And if Ludwig was not yet able to provide Marie a secure future, Anton was even further from marriage material. Although he came from a well-to-do Viennese medical family, he was only in his second semester of medical school.

And yet when Ludwig and his family also arrived at the holiday destination, the contrast between the two men struck Marie with painful clarity. She later acknowledged that in Ludwig, "a certain feebleness and weakness of will had bothered me all along."[2] With her feelings for him so diminished and no obvious way to extricate herself from the promise of marriage, she sank into despondency. Ludwig sensed her withdrawal and tried to make sense of the change. One day, when he and his mother were about to confront the unhappy young woman about her intentions, it all became too much—Marie escaped from the room and "ran into the forest, up the mountain, until she could no longer."[3] After that summer, the two muddled on with Marie feeling increasingly distant and Ludwig still hopeful that they might overcome whatever had come between them. Marie felt trapped.

The following spring, her oldest brother, Adolf, came home to Vienna for the Easter holiday. He had just started a professorship of jurisprudence at the University of Zurich and now invited Marie to join him there for a visit. She gladly accepted the invitation to escape her increasingly oppressive ties to Ludwig. On the train to Switzerland, she finally broke down. She confessed to her brother that she could no longer get herself to love Ludwig and that she had fallen for their friend Anton. Her brother took the sobbing girl into his arms and comforted her as best he could: "You've just grown apart. You're no longer suited for each other, nobody is to blame." Marie found much-needed solace in her brother's support and finally composed a letter to break off the engagement after her arrival in Zurich. She later remembered feeling "like someone who was finally free after a terrible incarceration."[4]

Adolf would remain in Zurich for three years. His youngest brother, Franz Serafin, soon joined him there to study physics at the newly founded Polytechnic University, later known as the Eidgenössische Technische Hochschule (or ETH). During this time, Marie came for a second extended visit, and the siblings imbibed the rich intellectual life of the city. Through Adolf's juridical and Serafin's scientific colleagues, they became part of a lively circle. Among their friends from this time, they counted the famous architect Gottfried Semper, the archeologist Karl Dilthey, and the well-known poet, writer, and politician Gottfried Keller.[5] Though

Keller had already turned fifty by the time he met Adolf, a warm friendship developed between the two, and he would carry on a charming correspondence with Adolf and especially Marie until shortly before his death in 1890.[6] The writer also joined the Exners over the summer of 1873 in the Salzkammergut and again in 1874 when he visited them for three weeks in Vienna.[7]

By the time of Keller's second visit with the Exners, much had changed for the siblings. Adolf had left Zurich for a position at the University of Vienna, and he and Marie were now sharing part of a house in the city. They had prepared a room for Keller's visit, "completely quiet and friendly with a rose bed in front of the window."[8] Marie's life, too, had moved in significant directions. Once she had freed herself from her ties to Ludwig, her relationship with Anton von Frisch blossomed. The two planned to marry in late fall of 1874. Anton had by then advanced to an assistantship under the famous Viennese surgeon Theodor Billroth. The same year he and Marie married, he was named professor of anatomy at the Art Academy (Akademie der Bildenden Künste), a position he would hold for the next twenty-five years. His career would continue to flourish. He published widely on topics that ranged from bacteriology and diagnostics to surgery and emerged as a central figure in the establishment of the modern discipline of urology.[9] Anton's professional success ensured that the von Frisches enjoyed a high standard of living. While Karl would later remember them as never having been overly wealthy, he acknowledged that his upbringing was privileged, with the family's social and intellectual lives ably facilitated by Marie.

Marie was also instrumental in the purchase and running of what would become the von Frisch family's permanent summer residence and safe haven during the world wars, as well as the site for many of Karl's most important experiments.[10] Shortly after she and Anton married, they acquired an old mill in Brunnwinkl nestled by the picturesque Wolfgangsee in Austria's Salzkammergut where they had first met. Gradually the family acquired more property surrounding the mill, and eventually their country residence included five houses around the lake. The houses became

FIGURE 1.1. The mill and surrounding houses in Brunnwinkl, Austria, that Marie von Frisch convinced her husband, Anton, to purchase in the 1870s. In the decades to come, the five houses on Lake Wolfgang would become a vibrant intellectual home for the von Frisch family and their friends. The summer colony would also offer family and friends refuge during the two world wars and serve as an important site for Karl von Frisch's scientific work. (Nachlaß Karl von Frisch, Bayerische Staatsbibliothek, Munich, ANA 540.)

an enduring vacation spot, not just for the von Frisches but also for their varied and influential friends.[11] In this setting, Marie ran a sort of country salon—the von Frisches' guests included Marie's own learned family as well as the likes of Theodor Billroth, the composer Johannes Brahms, and the Austrian writer Marie von Ebner-Eschenbach, who read to the Brunnwinklers from her manuscript in progress.

The von Frisches' own family would also expand over the next decade and a half. The summer after Anton and Marie married, they welcomed their first son, Hans. A little less than two and three years later, Marie gave birth to their second and third sons, Otto and Ernst. The youngest, Karl, was born after a gap of eight years in November of 1886. As the youngest of the family, he grew up surrounded by nature and animals as well as the intellectual guidance of Marie and her brothers, who all held university professorships.

During his childhood, Karl planted deep personal and intellectual roots in the soil around the mill in Brunnwinkl. It was there that he performed his first research experiments on eels and started a collection of local fauna and flora, which over the years swelled to almost five thousand specimens. He was captivated by the abundance of natural life and recalled having "wanted to collect everything, not only, say, butterflies or any other selected group as most people do." It was also at Brunnwinkl that he delivered his first "scientific talk" to an audience that included his uncle Sigmund Exner, who would later become his physiology teacher and intellectual mentor at the University of Vienna.[12]

Marie also encouraged her son's interest in animals. When she traveled to the coast of Istria to recover from an illness, she took the boy with her. Many years later, von Frisch recalled how he had lain "for hours between the cliffs, motionless, watching the living things I could see on and between the slimy green stones just below the surface of the water. I discovered that miraculous worlds may reveal themselves to a patient observer where the casual passer-by sees nothing at all."[13]

FIGURE 1.2. Karl as a toddler in Vienna surrounded by toys, circa 1888. (Nachlaß Karl von Frisch, Bayerische Staatsbibliothek, Munich, ANA 540.)

FIGURE 1.3. The family string quartet comprising Karl (*far right*)
and his older brothers, Otto, Ernst, and Hans (*left to right*).
(Nachlaß Karl von Frisch, Bayerische Staatsbibliothek,
Munich, ANA 540.)

Von Frisch began his studies with private tutors at home. Later he enrolled in the exclusive humanist Schottengymnasium in Vienna, where his future fellow Nobel laureate, Konrad Lorenz, would study some seventeen years later. Formal study was not an unambiguous pleasure and success for the young von Frisch, and he was especially tormented by the more theoretical subjects, such as Latin and mathematics.[14] But his passion for all things living made him a diligent and able student of natural history and biology. These became his favorite subjects and allowed him to draw connections between the knowledge of books and what he encountered outdoors as well as the many animals he kept as pets.

By the time Karl entered secondary school, his home menagerie had swelled to impressive proportions. He had accumulated an astonishing 123 different species of animals, of which only 9 were mammals.[15] Von Frisch was also an accomplished fresh- and saltwater aquarist, an experience that he later credited with having trained him in the art of careful observation. But he developed perhaps the most intimate connection with a bird: a small Brazilian parakeet named Tschocki that lived with the family for some fifteen years. The bird made a deep impression on the boy. He recalled how it preferred him to all other members of the family. Tschocki was his "constant companion at home, sitting upon my shoulder, dozing in my lap, nibbling at the papers and pencils on my desk, or otherwise engaged about my person. . . . At night he slept next to my bed, and first thing every morning I would reach into his cage to pick him up and talk to him." His mother had served as an example in bird care—every year she would hand-rear a blue tit she bought from a pet store and then release it come spring.[16]

FIGURE 1.4. Karl with his uncle, the experimental physiologist Sigmund (Schiga) Exner, during his student days. This photo was taken in Brunnwinkl and shows Karl's natural history collection. As a young boy, Karl had started to collect the specimens native to the region surrounding the mill, and parts of his collection are still on view nearby at the town museum of St. Gilgen. (Nachlaß Karl von Frisch, Bayerische Staatsbibliothek, Munich, ANA 540.)

In 1905, Karl von Frisch entered the University of Vienna medical school. At the time, Vienna had emerged as a bustling metropolis—from the composer Gustav Mahler to the architect Adolf Loos to Gustav Klimt and Sigmund Freud, the city teemed with the period's most important thinkers and cultural innovators. While this vibrant capital of the Austro-Hungarian Empire was neither the only nor, indeed, the first city in which these currents took hold, "European modernism reached its very purest and most concentrated expression in Vienna at the turn of the century."[17] Natural science, too, was part of this flowering, and the Exners—Karl's mother's family—counted among the city's most illustrious intellectual families.

From the early nineteenth century through the Second World War, three generations of Exners would excel in physics, physiology, meteorology, avant-garde art, law, and medicine, and the family claimed no fewer than ten university professors. They were as intimately entwined in the intellectual currents of the city as they were to the growing summer colony in Brunnwinkl, with their private and professional lives seamlessly blending through work, leisure, and marriage. Together with others, including the physicists Ernst Mach and Ludwig Boltzman, the Exners strove to supply a liberal rejoinder to the specters of left-wing anarchy and right-wing clericalism that beset Austrian politics and culture by offering a probabilistic approach to science. "They 'tamed' uncertainty by quantifying it," in the words of the historian of science Deborah Coen.[18]

But despite the growing intellectual and professional successes that surrounded him, young Karl felt far from sanguine about his own path at the University of Vienna. His heart was in zoology, but his physician father had convinced him that medicine would ensure a more certain future. These misgivings notwithstanding, his time at the university was not spent in vain. He acquired a firm grounding in anatomy, zoology, and histology. Most important, he worked with his uncle, the aforementioned experimental physiologist, Sigmund Exner. Exner had studied medicine at the University of Vienna under Ernst Wilhelm von Brücke and later at Heidelberg under Hermann von Helmholtz, two of the foremost physiologists of the nineteenth century.

The experimental rigor and elegant demonstrations Exner brought to his work deeply impressed the young von Frisch. Many years later, he would recall his uncle's course with enthusiasm. Exner had "conveyed the function of human organs with exemplary clarity and free of unnecessary baggage." And he praised his uncle for his ability to "compellingly support his claims with well thought-out demonstrations." Each night, von Frisch would recopy the notes from Sigmund's lectures and carefully study them. He was also "burningly interested in the physiological [laboratory] exercises" that accompanied his studies.[19] In addition to these formal requirements, von Frisch also took up his first independent research project under Exner's guidance. In these experiments, he expanded on his uncle's own investigations of compound eyes to study the pigments in the eyes of moths, lobsters, and shrimp.[20]

In a vertebrate's eye, the iris changes shape to regulate the amount of light that passes through the pupil to the retina. Invertebrate eyes lack this structure and instead depend on moving pigments to regulate light influx. Von Frisch set out to uncover the nature and cause of these pigment shifts. To this end, he moved invertebrate animals from dark to light and vice versa and then examined their eyes. The two different pigment configurations that resulted from these differing light conditions he termed "light eyes" and "dark eyes." He performed these examinations on both living animals in various stages of light and darkness and animals he killed at regular time intervals in the light-dark transition. In a subsequent set of experiments, he exposed living animals to lights of different wavelengths and determined that the pigments were most sensitive to blue-violet and violet light.

After studying the speed of pigment transitions, he went on to examine whether light itself or nervous stimulation triggered the pigment changes. He tested all the usual suspects for nervous stimulation—electrical stimuli, acids and bases, different temperatures, and even radium and X-ray radiation—but found that none of them caused a change in the animals' eye pigments.

When he published the results, he opened with a disclaimer: "The following experiments by no means led to clear results. If they are published nonetheless, it will be because in this case the negative and often inexplicable results appear to be not without interest and perhaps will stimulate further investigations."[21] And indeed, the thorough and careful nature of the study showed a promising and diligent student of comparative physiology. Von Frisch systematically sought to isolate and examine the various factors that might be at play in the pigment changes of the eye. He took his uncle's work as a touch point, revising and expanding on his findings. The style is very clearly that of an experimental physiologist who studied animal organs in vivo and in vitro.

In addition to dutifully reporting the various pitfalls and failures of the experiment, the paper also gives a vivid sense of what it was like to work with these animals as the young scientist was learning to perform delicate surgeries. In a group of shrimp, for example, he severed the hairlike fibers of the animals' optic nerves. The text is not always clear as to whether the animals survived the procedures, or indeed, whether this was desired for the experiment. But we can glean that at least in some cases, he performed experiments with postoperative animals that had evidently survived the earlier procedures.

Even when von Frisch was not wielding a scalpel, manipulations of the animals required considerable skill and patience. To test the effects of partial illumination, he smeared a tacky alcohol-soot mixture on the animals' eyes. He described the procedure in some detail and deemed it "not entirely easy," as "the shape, smooth nature, and great sensitivity [of the eyes] to even the smallest amounts of light are challenging circumstances." The effort was made even more difficult by the animals' active resistance to the treatment: "Because the animals know to deftly remove the annoying cap, one also has to include the eyestalk." "After several failed experiments," he settled on a combination of soot, celluloid, and ethanol that made a "quick-drying paste, and with this the eyes are pasted over."[22]

Although von Frisch in his scientific publication maintained a degree of detachment befitting a budding scientist of the time, his later reflections on these experiments tell a somewhat different story. He recalled how, although he "went to work with great enthusiasm," he "was very soon faced with a serious conflict. [He] had to stimulate the eyes of living crustaceans with electric currents, a manipulation which clearly was to them a most unpleasant experience." His reluctance was so strong that he wrote how "every single experiment cost me an effort. Though finally my scientific zeal got the better of my compassion, I believe that even in later years I could not have brought myself to undertake similar research on birds or mammals with their more highly developed and presumably more sensitive nervous systems." Von Frisch's sympathies were visceral but also strongly mediated by his scientific understanding of kinship and an evolutionary hierarchy that placed humans and mammals at the top of the pain-perceiving scale. And although the study was ultimately inconclusive, he later recognized that the work awoke in him an abiding interest in experimentation and sensory physiology.[23]

In 1908, this growing passion for experimental zoology would lead him away from the University of Vienna. After passing his medical exams with distinction in all subjects, he decided that he no longer wanted to endure the medical curriculum, which had become increasingly clinical. He left medicine and Vienna for Munich to pursue a doctorate in zoology at Richard Hertwig's Zoological Institute. Hertwig had studied at Jena, during the heady time of Ernst Haeckel and Carl Gegenbaur's joint reign. The two had provided students with a rich mix of anatomy, zoology, and evolution. Richard and his brother Oscar were two of their most promising disciples. Although this was the first time von Frisch was to meet Richard Hertwig in person, as a boy he had already read a book on anemones by the Hertwig brothers.[24]

Hertwig's institute was located in an old monastery in the very heart of Munich. True to its roots, the monastic quarters translated into an intensely focused setting for von Frisch's studies. He found that in this "harmonious isolation," it was "not difficult to concentrate on one's tasks." The institute was a center of a new kind of science where "next to morphology and comparative anatomy, experimental zoology came to life."[25] Von Frisch became an enthusiastic visitor of Hertwig's large practicum, a combined lecture and laboratory course that introduced students to the foundations

**FIGURE 1.5.** A view into the courtyard of the former monastery that was Hertwig's (and later von Frisch's) Zoological Institute in Munich. The twin towers of the Frauen-kirche are visible in the background. (Nachlaß Karl von Frisch, Bayerische Staatsbibliothek, Munich, ANA 540.)

of the new biology. The class met six times a week at the bracing hour of 7:00 a.m.

It was during this course that von Frisch would make the acquaintance of another new student at the institute, Otto Koehler, who would become a close friend and ally. In their early days in Munich, the two soon fell into a nightly ritual. After they completed their work at the institute, they would ride their bikes back to von Frisch's student apartment. Over simple dinners of bread and butter, sausage, and cheese, they chatted freely about the day's events, their plans, and whatever else was on their minds. An easy friendship developed despite the young men's divergent backgrounds. In addition to their respective Viennese and East Prussian dialects that at times made it comically difficult to understand each other, Koehler had experienced a very different upbringing from that of von Frisch. Orphaned at an early age, he was alone and without immediate familial or financial support. After one of his visits to the von Frisches' summer residence in Brunnwinkl, he expressed his admiration and longing for the close-knit and intellectually vibrant family. Von Frisch later remembered the exchange as having awakened in him an appreciation for how special his own upbringing had been.[26] But as much as Koehler lacked in domestic riches, he threw himself into learning; already as a young man, he was deeply knowledgeable about history, art, and especially music. Von Frisch too had enjoyed a humanistic education and studied the violin from an early age to complete a string quartet with his three older brothers. But compared to Koehler's accomplishments, he dismissed his own as dilettantish.

In addition to Koehler, von Frisch also became close to some of the period's most promising young scientists, including the geneticist Richard Goldschmidt and the zoologist Franz Doflein. Goldschmidt served as Hertwig's chief assistant, and Doflein had just been appointed assistant professor of systematics and curator of the Zoological Museum. Von Frisch would later remember fondly the nature excursions he undertook with Doflein in and around Munich and to more remote destinations.[27]

But in 1909, after only two semesters in Munich, von Frisch returned to Vienna to be closer to his family while reasearching and writing his dissertation. Back in Vienna, he approached Hans Przibram, who led a biological research facility near the Prater amusement park on the outskirts of Vienna. The Vivarium, or Biological Research Station (Biologische Versuchsanstalt, BVA for short) as it was officially called, was founded in 1903, only six years prior to von Frisch's arrival. The neo-Renaissance building had been built for the 1873 Viennese World's Fair and was purchased in the early twentieth century by Hans Przibram and two of his colleagues, Wilhelm Figdor and Leopold von Portheim, through their private monies.[28] Turn-of-the-century Vienna would come to be remembered as the home of Freud and Klimt, Mahler and members of the Secession.

The natural sciences at the University of Vienna also enjoyed a high point around 1900. And yet some disciplines fared decidedly less well in this setting. Biology, in particular, was underfunded and clung to a traditional approach. Matters were not helped by the fact that keeping live animals in the new science building at the university was forbidden. These factors made it virtually impossible to pursue the new experimental biology with which Przibram had become enamored.[29] Flush with familial wealth and highly capable and trained, Przibram and his colleagues dedicated their energies toward building an institute to house the new kind of highly experimental and interdisciplinary life science. The work at the BVA, according to Przibram, "should not limit itself to only certain problems but should rather pull into its domain all large questions of biology." He went on to write expansively about the organisms that would be called upon to serve these ends: "Animals, plants, inhabitants of fresh water, the seas as well as terrestrial parts are all equally welcome."[30]

While much of the nineteenth century had been dominated by studies of morphology of dead and often extinct animals, the second half of the nineteenth century saw the ascendance of a new kind of life science, one that turned its attentions to living processes such as inheritance, regeneration, and reproduction. Its practitioners increasingly looked to the experimental and quantitative methods of the physical sciences and physiology as models. The BVA was to offer an interdisciplinary setting with four departments—zoology, botany, physical chemistry, and physiology—dedicated to the problems and processes of life. To succeed in these pursuits, it became critically important that the animals and plants under investigation endured the probings by scientists long enough to reveal their behaviors and structures over time and across generations. The BVA was designed specifically to meet these needs.

Przibram and his colleagues had completely gutted the place and installed state-of-the-art equipment. Fresh- and saltwater tanks with elaborate aeration and circulation systems housed aquatic animals (and the latter were filled with seawater brought in by railcar from Trieste, Italy). Terraria, cages, gardens, outdoor ponds, pens, and stables all converged on keeping alive the living materials of science. In addition, high-moisture rooms mimicked tropical environments and a cave built five meters belowground ensured absolute darkness and a steady 12 degrees Celsius. Special contraptions served to alter conditions of gravity on animals and plants, and dark rooms allowed researchers to carefully regulate ambient lighting.[31] Years after it had been taken over by the Nazi state and destroyed in the war, von Frisch remembered the BVA approvingly: "There it did not smell of clove oil and denatured alcohol, there reigned the living animal—and that's where I felt myself drawn."[32]

When Karl von Frisch first arrived, Przibram assigned him a study of the developmental history of the praying mantis, an organism on which Przibram himself had worked. But von Frisch was soon bored with the topic and happened to observe a fellow student working on pigmentation changes in minnows. The animals underwent rapid color changes when swimming in differently lighted environments. Von Frisch was hooked.

He felt the work was a natural outgrowth of his earlier study under Exner on the pigment changes in compound eyes of invertebrates. Przibram gave his blessing, and von Frisch promptly changed his topic to how these color changes were regulated in fish. Notably, to von Frisch's own mind, he had once again "fallen for a comparative physiological project," and he resumed his visits to his uncle Sigmund at the Physiological Institute for advice.[33]

Departing from an earlier study by Félix Pouchet, von Frisch began by cutting the sympathetic trunk in the fish's body just below its dorsal fin. Almost immediately the fish's tail turned dark, starting from the point of the incision and moving down its body. Afterward, when he stimulated the fish's sympathetic trunk with electricity, its skin once again turned light, suggesting that contraction caused the cells to appear lighter. Conversely, the relaxation prompted by the cutting of the nerves caused the pigments to move to the skin surface, thereby giving the darker appearance. When von Frisch gradually moved the point of incision in different fish to successively higher positions along the spinal cord, more and

**FIGURE 1.6.** Images of fish from von Frisch's early work on the nervous control of pigment changes. (Karl von Frisch, "Beiträge zur Physiologie der Pigmentzellen in der Fischhaut," in *Festschrift zum sechzigsten Geburtstag Richard Hertwigs*, vol. 3 [Jena: G. Fischer, 1910], plate 7.)

more of the body changed color posterior to the cut. That is, until he reached a point just above the dorsal fin—suddenly the head turned black while the posterior part below the incision remained unchanged. Severing part of the optic nerve at the source behind the roof of the eye socket also caused the head to turn dark from eye to snout. Further experiments with the animals revealed that the contraction of pigment cells was mediated by two separate nervous pathways—one was responsible for the changes in the anterior part of the body, while the other controlled those posterior to the fifteenth vertebra.[34]

After only one semester, von Frisch submitted his thesis and stood for his oral exams. His major and minor examination fields were in zoology and botany, respectively. An additional field was required in philosophy, a subject about which von Frisch admitted to knowing "next to nothing." Counter to expectations, however, he fumbled through the zoological part of the exam, which should have been his strongest subject. His examiner had asked him questions about what he considered "very boring matters" and for which his time in Munich had ill prepared him. But to his even greater shock, he performed with "distinction" in the philosophical part of the exam. This surprising outcome coincided with an opportunity to vent some frustrations he had been harboring over his examiner's rejection of evolution. His "first and only question" was about the theory. Von Frisch apparently forgot himself and held forth on its validity with some fervor for the remainder of the examination period. The professor, duly impressed, perhaps as much with the young man's passion as with his knowledge, awarded him distinction on the grounds that he had "expressed [his] own opinion."[35] The incident points to an important and enduring conviction von Frisch held—namely, that functional traits in organisms are adaptive and that structures bear witness to their evolutionary histories.

In 1910, von Frisch returned to Munich to assume an assistantship under Hertwig. He now worked closely with Hertwig's first assistant, Richard Goldschmidt, and helped run the Zoological Institute's large laboratory course in which he had met his friend Koehler just two years prior. In addition, he was also able to pursue his own research interests. And it was through his project on the sensory physiology of fish and later bees that the young researcher would soon become embroiled in one of the most heated debates of his career.

# The Bees That Could

⚜

In a letter to his mother written on October 22, 1912, Karl von Frisch drew a little face of a man with a narrow, downturned mouth and a thin mustache of considerable wingspan. Next to the doodle, he regretted that "it did not quite succeed." The comment was about the likeness the caricature was supposed to capture—Carl von Hess, the director of the prestigious Munich Eye Clinic. In real life, von Frisch explained in the letter, the man's visage appeared "sharp and egotistical." This matched the stern "enter" he had barked through the office door "like a commanding major" in response to the twenty-six-year-old von Frisch's knock. Although the meeting with the elder colleague unfolded much less tensely than von Frisch had expected (von Hess had "inquired after Hertwig and family circumstances" and even seemed "charming"), the encounter took place in the midst of a protracted dispute between the two that would continue even after their in-person conversation. Twenty-six years von Frisch's senior, von Hess proved a formidable opponent, and years later von Frisch would remember their tangle as one of the formative experiences of his career.

Those early years back at the Zoological Institute in Munich as Hertwig's assistant proved extraordinarily fruitful for von Frisch. He again frequently conferred with his uncle Sigmund (Schiga) Exner. In a letter to his father, he recalled how "Uncle Schiga" had visited him one morning at the Zoological Institute. He wrote of how his uncle had the ability to help him sort and clarify his thinking, like a "soup strainer on my flabby thoughts." After another meeting, he recalled having learned "more in that hour than from Hertwig the entire last year." During this time, von Frisch would also see Otto

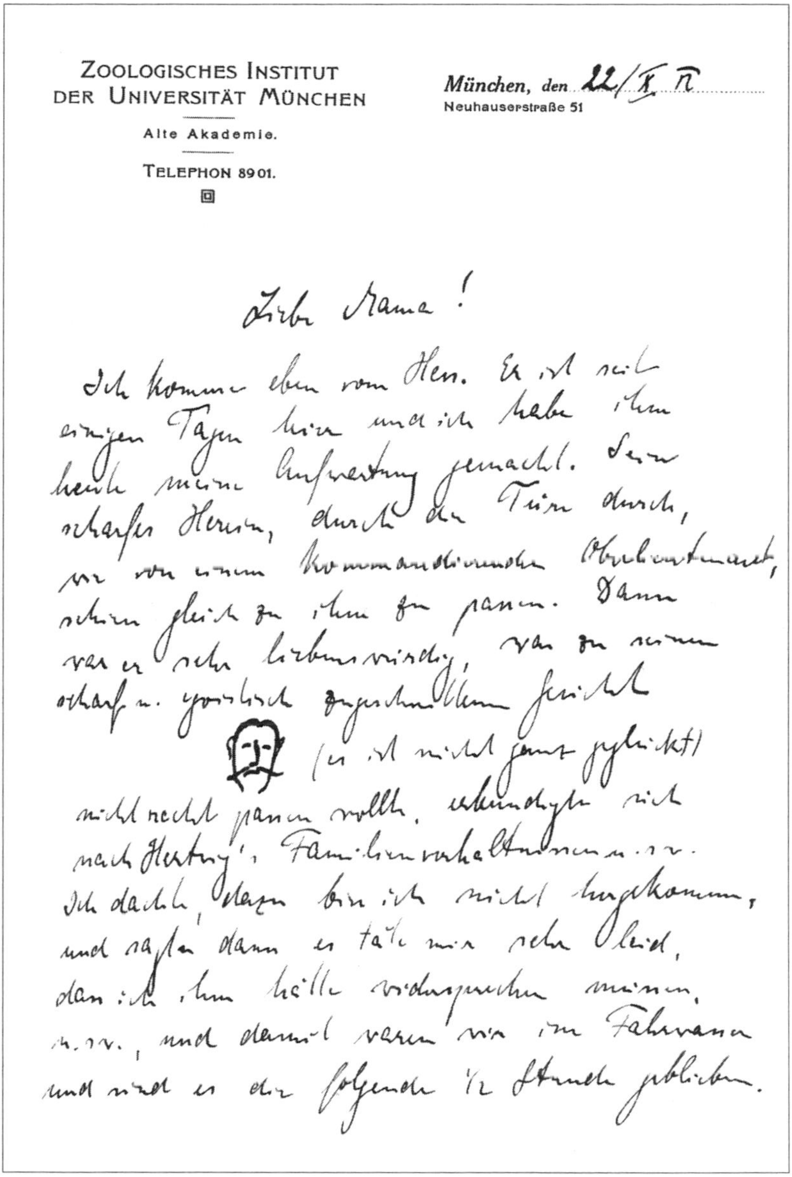

*Liebe Mama!*

*[handwritten letter text]*

FIGURE 2.1. The letter by von Frisch to his mother written on October 22, 1912. The little cartoon was of his nemesis, the ophthalmologist Carl von Hess, with whom he tangled over whether fish and bees were capable of seeing in color. (Nachlaß Karl von Frisch, Bayerische Staatsbibliothek, Munich, ANA 540.)

Koehler again frequently. The two visited Italy in the summer of 1912 to study marine fauna and travel together. Von Frisch quipped to his father that he and his friend had been asked to dinner at Hertwig's house and were "treated like a married couple when it comes to invitations."[1] Alongside these deepening ties to friends and colleagues in the years leading up to the First World War, the dispute with von Hess dominated much of his professional life.

The controversy focused on whether fish and invertebrates possessed color vision. Von Hess had performed extensive investigations into the abilities of various animals to see in color. He concluded that fish as well as all invertebrates do not possess this capacity, much like humans with color blindness. Von Hess had based his conclusion on work he had performed at the Naples Zoological Station on the Mediterranean sand smelt (*Atherina hepsetus*), a slender-bodied fish abundant in the local waters. He gratefully acknowledged "the gentlemen of the station" who supplied him "over several weeks almost every second to third day with often hundreds or more of the suitable little fish for the experiments, so that my observations were based on over a thousand fish." The immature animals were known to be positively phototactic; that is, they move from dark to light areas in their environment. Von Hess released several dozen fish at a time into a basin and then shone a light through a prism onto the tank. The prism resolved the white light into its characteristic rainbow of colors and projected them along the side of the tank. He then had a helper from the Zoological Station take a photograph to capture the positions of the small fish, so he could later trace the animals to capture where they had congregated during the trial. He reported, "Almost all the animals began immediately to swim in the direction of the yellow-green to green of the spectrum and within a few seconds the majority of the fish had congregated in the area." To von Hess, this finding suggested that the animals saw that area (around yellow and green) as the lightest, since the animals tended to swim toward light. Von Hess noted that this corresponded to how color-blind humans (as well as humans whose eyes have adjusted to low-light conditions) perceive the color spectrum: "The fish behaved in all our investigations close to or exactly as would totally colorblind humans in all light conditions, and normal, dark-adapted humans in dim lighting, if they were asked to identify the area that appeared to them the lightest."[2] He concluded that the animals responded to differences in brightness but were incapable of distinguishing among hues.

Von Frisch was skeptical. His dissertation work on the pigment changes in fish had led him to believe that the animals were closely attuned to the colors of their surroundings, and von Hess's claim that fish did not perceive colors prompted him to take up the question himself. Von Frisch first publicly challenged the elder colleague's findings at a gathering of the German Zoological Society in Basel the year before their in-person meeting. The paper began on a friendly enough tone, with von Frisch reminding his audience of "the cleverly designed and precisely conducted comparative experiments on the light sense of animals by the ophthalmologist Hess."[3]

In his own experiments, von Frisch used a fish species known as corkwing or goldmaid (*Crenilabrus melops*) that changes its coloring to mimic its surroundings. The species is common to the waters around Naples, where he too had conducted the work while at the famous zoological station. For his experiment, he kept three groups of brightly colored fish in monochromatic light conditions. He wanted to test whether the animals only responded to gradations in brightness, as von Hess had claimed, or were sensitive to the actual color of their surroundings. To investigate the question, he made a slight but significant adjustment to his colleague's experimental scheme. Rather than relying on how the fish moved through their tank (that is, toward or away from a given light source), he decided to use their physical color changes as indicators of their ability to perceive and differentiate light. The logic was that if the animals could adjust the pigmentation of their bodies to blend into their surroundings, then they must somehow be able to perceive the features of the environment to which they matched their coloring.

For the first part of the experiment, von Frisch kept one group in a tank with green light, another in red, and a third in daylight as a control. He found that the fish retained their coloring even after weeks spent in the monochromatic conditions: the animals were shaded green, red, and tan, respectively. To von Frisch's mind, the fact that these animals maintained their different coloring over long stretches of time suggested that they must be able to perceive the color of their environments not just with respect to relative light-dark values. "By what means," he asked rhetorically, "should these fish that were held over weeks in monochromatic light, recognize that they are in red or green light, if they do not see it as qualitatively different?"[4]

In subsequent experiments, von Frisch returned to minnows, the species he'd used in his dissertation on the nervous mechanisms that control pigmentation changes. Departing from his earlier studies, he now focused on the fact that the animals had two distinct mechanisms for changing their coloring. The first was a light-dark adjustment caused by the contraction and relaxation of black pigment cells. Second, the fish were also able to change their overall coloring through contraction and relaxation of red and yellow pigment cells that cluster along their bodies. Significantly, the light-dark and color pigment changes occurred at different rates— the black pigment cells responded in a matter of seconds, while the color adjustments occurred over the course of hours. This meant that the two mechanisms could be observed as distinct events, and it was on this feature that von Frisch would base his argument.

In this part of the experiment, he selected from a larger group of minnows two specimens that appeared identically colored when exposed to backgrounds of different colors and light-dark gradations. For the remainder of the experiment, he assumed that these two animals were equivalent in their reactions and could be used interchangeably as well as side by side for comparison. He next placed the animals in two separate tanks and exposed one of them to a sheet of yellow paper while the other was tested against various shades of gray. The goal was to find the piece of gray paper that made the one fish appear the same shade as its fellow in the yellow environment. This would suggest that the yellow and gray sheets were of the same light value (or brightness) despite being of different colors (or hues). Next, von Frisch observed one of the animals slowly undergoing further changes in its coloring. After a little more than a half hour, the animal on the yellow surface started to change to a more yellow color while the animal on the gray surface remained unchanged. He ascribed this second transition to shifts in the red and yellow pigment cells, which occur more slowly than the black pigment adjustments. To von Frisch, this second color change demonstrated that the fish perceived a qualitative difference between the two equally bright environments.[5]

Although the findings seemed to contradict von Hess's claims that fish were unable to perceive hues, von Frisch ended on a solicitous note: "For the time being, these results of course only speak to the animals on which they were obtained." Since he'd experimented on corkwing and minnows while von Hess's conclusions had rested on work with smelt, differences in their findings might be ascribable to sensory variance among the different species. He also offered that their work had been conducted under seemingly different ambient light conditions that might have caused the discrepancies between their findings. Or their differences might also be due to von Hess having experimented on juveniles, while von Frisch's conclusions had derived from work with mature animals. Maybe the animals' sensory physiology matured to color vision, he suggested. In the end, perhaps a compromise could be found: yes, their sense of relative brightness along the color spectrum might more closely match that of color-blind rather than normal human vision, but the fish also possess a sense of color.[6]

Despite von Frisch's professed agnosticism on the causes of their variant findings, von Hess did not take kindly to the challenge. In a 1912 publication, "Examinations on the Presence of a Color Sense in Fish," he delivered a pointed response.[7] While he granted that von Frisch's experiments were conducted "significantly more carefully" than those of other detractors, he denied that they offered "any support for assuming a sense of color in fish." "It is well known," he lectured, "that one of the first rules of color physiological experiments is to work only with matte (that is, nonglossy) pigment papers." To do otherwise was to introduce reflected ambient light, which confounded all experiments on color. Von Frisch had sinned in just this manner. And to von Hess's mind, the consequences of such a misstep were fatal to the work: *"Of all scientific claims, those drawn from experiments involving glossy papers are to be dismissed out of hand."*[8]

Von Hess also offered new experiments intended to question a more fundamental aspect of his young colleague's work—that a fish's body coloration was a reliable indicator of how the animal perceived its environment. This assumption, he admonished, had not been adequately tested: "As the physicist tests the measuring instrument as to its reliability prior to undertaking a measurement, the observer needs here above all to determine the smallest light differences of the environment that still cause a definitive change in the appearance of the fish."[9] Von Frisch, he argued, had failed to determine this increment and, therefore, to calibrate the fish as his "measuring instrument."

In response to this failure, von Hess had devised a special experimental setup that was to test the sensitivity of the fish's adjustments. An L-shaped tunnel with a light source at one end and a fish tank at the other was set up in a room with walls and experimental equipment that had all been painted matte black. The environment was designed to control for ambient and reflected light and to convey light of a "continuous and measurable, variable strength" to the animal's tank. Von Hess found that the fish's bodies did not change color reliably with respect to the brightness of the light beam. "Indeed, even when the environment was ten times brighter," the fish "weren't always lighter in their coloring" than the animals swimming on the darker surface. He reported having measured in different experiments that fish could perceive very slight differences in brightness. And yet these subtle differences in perception were apparently not expressed in the coloring of their bodies. "Fish don't adjust themselves nearly as precisely to their environments as von Frisch had assumed without conducting the definitive experiments," he declared. And therefore, "all conclusions von Frisch drew from his erroneous assumption ought to be disqualified." Von Hess closed with a summary dismissal of "all experiments conducted to date in an effort to show a color sense in fish" and noted that they "were without exception conducted without knowledge of color theory."[10]

Von Hess's implication that von Frisch was ignorant of color theory and had offered his critique from a lay perspective was highly provocative. Around the turn of the century, representatives of diverse disciplines (including physiology, psychology, and physics) weighed in on the issue of color, and whether it was a phenomenon best explained as having physical reality independent of the observer or better understood as residing in the physiology and psychology of the seer. In other words, should it be studied as a function of the outside world, or did the human eye and brain contribute essential elements to the equation? A heated debate between the physiologists Hermann von Helmholtz and Ewald Hering focused on this and other issues and captivated the attention of the wider scientific community.[11] Even more proximate to von Frisch, his uncles Sigmund Exner and the physicist Franz Serafin Exner had collaborated on an investigation on the topic just shortly before von Frisch took up the issue. In a joint publication of 1910, the Exners examined precisely the question of flower colors and insect visual perception and argued that the high degree of saturation of flower colors served to attract pollinators.[12]

The same year as von Hess's critique, von Frisch published a rejoinder titled "Are Fish Colorblind?"[13] Apropos the offending glossy paper, he now explained that he'd assumed it a nonissue, because in the side-by-side experiments, the gray and colored papers had both been glossy. The animals, he assured his readers, had been precisely monitored: they had swum for the same amount of time in identical glass containers and lighting conditions. "The only factor that was different for the two fish," he argued, "was indeed the color of the glossy paper."[14] Nonetheless, he reported having repeated the experiments using matte papers to satisfy von Hess's critique. The findings were the same. The color pigment cells of the fish always expanded on the yellow and red papers, while they contracted when the animals swam against black and white backgrounds.

He also addressed von Hess's most troubling objection—that the fish's physical coloration did not correspond closely to how the animals perceived their environments. Of von Hess's carefully designed light tunnel experiments, he objected that the colored light area to which the animals had been exposed was both too small and too far away from the fish to ensure a reliable reaction. He argued that to cause an adjustment in the animal's body pigments, the colored surface needed to make up a sufficiently large visual field for the fish. He also criticized the experiment for its unnatural setup: "Because of the abnormal conditions that captivity imposes on the fish, one cannot assume that all animals react as well as they might under natural conditions." [15] Here von Frisch could counter the ophthalmologist's claims of disciplinary superiority. A zoologist was surely bettered positioned to speak to the "naturalness" of an animal's environment than an ophthalmologist.

He concluded that only an "extreme skeptic" would persist in doubting these findings. "It would be gratifying if Hess now finally admitted that it has been proven that fish and color-blind humans perceive brightness equally and, at the same time, that fish have a sense of color." Von Hess, he urged polemically, might do well to "dedicate his considerable energies to solving the new problem of this apparent contradiction" instead of clinging to his assertion of total colorblindness in fish. [16]

In a subsequent publication of 1913, "Further Examinations on the Color Sense of Fish," von Frisch took a different approach to the question of fish vision. [17] This time, he no longer asked *whether* the animals could discern colors but rather *how* and *to what extent* they were able to do so. In other words, which colors exactly could they recognize and to what extent did these abilities differ from those of humans.

To answer these questions, he introduced a technique that was still relatively new in his experimental toolbox—conditioning the animals with food. He had already written about this type of work in a publication from the previous year in which he had adapted the method from an earlier work by von Hess. [18] In these publications, both scientists had fed the fish dyed meats and later tested their reactions to papers of the same color taped to the fish tank walls. Although von Frisch had taken care to match the colors of the test papers to those of the food dyes, he now decided to also use colored papers during training to ensure an exact match between the conditioning and test colors and to expand his testing palette beyond what the dyed meats had offered. To this end, he purchased standardized colors, as specified by Ewald Hering, one of the main interlocutors on color vision, whose laboratory had started to produce and sell standardized colored paper. Von Frisch also ordered a set of custom-made glass tubes, each of which had been fused to fully enclose a strip of Hering's colored paper. The top end of the tube was indented to form a slight dimple that could hold food. In this way, fish would see the colored paper while feeding, but the paper stayed dry despite being immersed in the tank. He also obtained fifty additional tubes that contained grays spanning from white to black. Each tube also contained metal beads opposite the food well to weigh it down and ensure it maintained a vertical orientation in water.

For the initial training phase, von Frisch placed a small amount of meat in the well of a tube with a red strip of paper. He next placed the tube alongside six others that contained gray papers but no food. He hung the tubes from a rack, which he then lowered into a tank that held a group of minnows. From time to time, he replenished the food and switched the position of the tubes along the rack to ensure that the fish did not associate the reward with a particular location. After a few weeks, he considered the animals "sufficiently trained," and the trial could begin.[19]

For the experiment, he again lowered a rack of six fresh glass tubes—one contained the same red paper as he'd used in the training phase while the others enclosed strips of different grays. This time, none of the tubes held food. At issue was whether the animals would recognize the color they had come to associate with food during the training phase. Von Frisch reported success: "the fish swam to the upper rim of the red tube and excitedly snapped at the glass. Indeed, frequently a fish stuck its head into the glass and searched the bottom of the food well . . . for the familiar food."[20]

He repeated these experiments with colors other than red and concluded that the animals were also capable of distinguishing yellow, green, and blue from grays. He further expanded the trials to test different colors against one another. Now a more complicated picture emerged—the fish, it turned out, were unable to distinguish between red and yellow, while blue and green appeared distinct from each other and the other colors. And with these findings, von Frisch not only demonstrated that the fish perceive color; he was able to elaborate the precise nature of their color vision. The work also featured an important methodological innovation— conditioning with food. The technique would become an essential tool in his repertoire, and much of his later work relied on it. While this is a well-known aspect of his work, less well known today is that it emerged as a consequence of (and indeed was adapted from) von Hess's work.

In addition to addressing methodological criticisms by von Hess, von Frisch also noted that the elder colleague had rejected some of his most fundamental observations. For example, von Hess had claimed that when he placed fish on yellow surfaces, the background color had had "no discernible effect on the coloring of the minnows."[21] Von Frisch recalled that "to my disappointment, von Hess was unwilling to come see my experiments that should not have succeeded based on his theory."[22] To combat this flat-out denial of his findings, von Frisch decided instead to enlist other witnesses: "On May 26, the gentlemen Prof. R. Goldschmidt and Dr. Buchner happened along by chance as the fish were on the same white background to assess their coloring. And without having been informed of the prior arrangements of the animals, they immediately recognized which of the fish had been on the yellow surface based on their coloring." Von Frisch also noted that he'd demonstrated the fish's color changes for four other colleagues, including his friend Otto Koehler. Even more significantly, he was also able to report his mentor's support: "Herr Professor Richard Hertwig very kindly agreed to undertake the experiments and record keeping with me; he confirms their correctness."[23] Although Hertwig sympathized with von Frisch on scientific grounds and agreed to publicly vouch for the experiment's accuracy, privately he also admonished him for the unseemly tone he had taken with the older colleague. And indeed, by 1913 the debate had reached a fevered pitch. In a letter to his father, von Frisch reported that

Hertwig had called the situation between him and von Hess "irreparable."[24] Nonetheless, Hertwig's public support sent a strong message—to deny these observations was not just to disagree with the still relatively unknown von Frisch but to call into question the experimental judgment and integrity of Hertwig himself, the director of the Zoological Institute. This willingness by Hertwig to vouch for von Frisch's work surely contributed to the latter's success in the dispute and beyond.

### Von Frisch Meets the Bees

Von Frisch's 1913 "Further Examinations on the Color Sense of Fish" was notable, not only for its thorough and innovative approach to color vision, but also for featuring an additional experimental organism—the honeybee. There, tucked among the growing data on how fish perceive and respond to colors, the insect made its first, brief appearance in the young physiologist's work. While fish were capable of distinguishing between red and black, he wrote, the same was not true of bees: "In this the fish's color sense deviates markedly from that of the honeybee, which confuses the same red with black."[25]

The comparison between fish and bees in the work was prompted by experiments he had begun in the summer of 1912. Like so much of his early work, he had come to the topic through von Hess. For von Hess, the animals provided little more than another data point among many. From mammals to birds to worms and insects, by this point, his comparative study of color vision had traversed much of the animal kingdom.[26] But for von Frisch, the move to the insects would prove consequential. When he began working on bees, he happened upon his ideal organisms. They were vastly complex in their sensory physiology and behaviors and amenable to being kept in Munich as well as at his country retreat in Brunnwinkl. Von Frisch's dedication to the insects would prove both fruitful and enduring for the rest of his career.

Von Frisch once again contrasted his own investigations with those of von Hess. As in his previous work, the elder colleague had relied on phototaxis in his experiments. Bees, too, are positively phototactic; that is, they move toward light. Von Hess surmised that the green and yellow-green portions of the color spectrum appeared brightest to the insects.[27] As was the case with fish, bees perceived that part of the spectrum as being lighter than all others (including yellow) and in this way matched the perception of color-blind humans. Thus, von Hess concluded, the animals too were color-blind. In his discussion of the results, von Hess acknowledged that this conclusion seemed to contradict much of what had been written about bees. Significantly, it flew in the face of the established wisdom on the relationship between flowers and bees that had been most famously elaborated by the naturalist and theologian Christian Konrad Sprengel.

In 1793, Sprengel had published *The Secret of Nature Revealed in the Structure and Pollination of Flowers*, which argued that the shapes, colors, and scents of flowers served to attract their specific pollinators.[28] In Sprengel's moral economy of nature, the scent of flowers served to attract bees, which then collected the nectar in exchange for pollination.[29] The work was still considered sound in its basic logic in the early twentieth century, albeit with the added heft of Darwin's theory of natural selection that gave the relationship between bees and flowers its evolutionary basis. As a committed Darwinist, von Frisch was sympathetic to this holistic view that closely linked the adaptations of flowers to their primary pollinators. Consequently, he again found himself skeptical of von Hess's rejection of Sprengel and decided to examine the bees' sense of color himself.

Von Frisch chose to pursue a line of research that differed from the phototactic approach favored by von Hess. At his country home, he set up a table "in a spot sheltered from rain and direct sunlight."[30] He had exposed sheets of photosensitive paper to light for increasing amounts of time to create a series of sheets of progressively darker shades. In addition to these grayscale papers, he also used two different shades of yellow. He duly noted that all papers were matte and had been purchased from the Richard Nendel Company in Leipzig, another supplier of standardized, colored paper.

**FIGURE 2.2.** The frontispiece to Christian Konrad Sprengel's *Das entdeckte Geheimniss der Natur im Bau und in der Befruchtung der Blumen* (The revealed secret of nature in the structure and pollination of flowers) published in 1793. The book and its gorgeous plates of flowers and insects argued that the colors and scents of flowers serve to attract their pollinators, especially bees. Charles Darwin cited Sprengel in *On the Origin of Species*, and von Frisch explicitly aligned his own arguments with the late eighteenth-century work. (Special Collections Research Center, University of Chicago Library.)

For the training phase, von Frisch arranged the gray and yellow papers on the table in random order. On each paper, he placed a glass dish. Of these, he filled only the two dishes on the yellow papers with sugar solution and left the others empty. He now lured bees from a nearby hive by smearing a large sheet of cardboard with honey and walking it slowly to the table so the bees could follow him. Once the insects were led to the experimental setup, they began to feed from the two sugar dishes on yellow. He now disposed of the honey-covered sheets, as sugar solution proved easier to handle; its concentrations could be carefully controlled and, unlike honey, it contributed no potentially confounding odors to the experiment. "Soon there developed a lively traffic of bees," he reported.[31] At the same time, he switched the location of the yellow papers from time to time to ensure that the animals were trained to color rather than location. After feeding the insects for two days on the yellow squares, he was ready to begin the experimental trial.

He again arranged a series of fresh gray and two yellow papers. But now he filled all the dishes with sugar water including the ones on gray. For the next ten minutes, he and a group of helpers observed and counted the number of bees that alighted on each of the papers. The goal was to see whether the insects would preferentially feed on the yellow over the gray dishes. He reasoned that "if bees behave like completely color-blind [people] in their faculty of sight, that is, if yellow has no color value but only brightness for them, then under these circumstances one expects that they would confuse yellow with certain gray papers, indeed, with those which for them exhibit the same degree of brightness as the yellow."[32] But, he added, "that was not the case." Rather, the "bees . . . continued to fly to the two yellow papers and crowded the sugar water on them." In contrast, the gray papers "remained ignored." Repeating the experiments with blue instead of yellow papers among the array of grays yielded similar results—during the experiment, the blue-trained bees showed a clear preference for the color over grays. Thus, von Frisch concluded that the animals also recognized blue and yellow by something other than their relative brightness.

In a subsequent set of experiments, he tested different colors

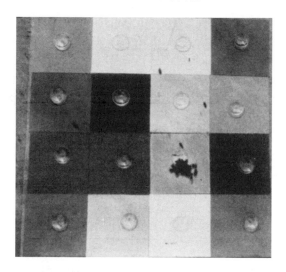

FIGURE 2.3. A photograph of what would become one of von Frisch's most famous experiments. Cardboard squares of different shades of gray were arranged in a random checkerboard with one colored square interspersed. The color of the single square—blue in this case—matched the color on which bees had been fed during a training phase. In the image, no food is present, but the black specks on the (only blue) square positioned second from the right and bottom are bees that congregated in search of food. The presence of the bees on the blue square suggested to von Frisch that the insects were capable of recognizing blue by hue rather than brightness. (Karl von Frisch, *Der Farben-sinn und Formensinn der Biene* [Jena: Gustav Fischer, 1914], plate 1, fig. 4.)

against one another. When he exposed a group of bees that had been trained to yellow to an array of colors (rather than shades of gray), he found that the animals clearly recognized the yellow to which they had been trained. However, their preference for that color was not exclusive: while most of the bees alighted on the yellow, several also landed on orange and yellow-orange papers. Comparing the bees' discernment to that of humans, von Frisch observed that these colors are also "to our eyes the most similar." At the same time, the yellow-trained bees "totally avoided the red, green, blue, and purple papers," suggesting that those shades must have appeared to them very different from the yellow, orange, and yellow-green shades.[33]

In further trials using animals trained to purple, von Frisch again observed a preference for the training color as well as a few additional hues. This time, the animals flew to the purple and blue papers. At the same time, the animals avoided all yellow, red, and green shades. Von Frisch also reported that all attempts to train the bees to red had failed; the bees alighted in large numbers on the darker shades of gray and black and showed no preference whatsoever for red over these darker shades. The result suggested that the animals were incapable of seeing red and recognized the paper only by its light content, as von Hess had insisted of the animals' vision in general. Ultimately, von Frisch offered an elegant explanation of the apparent "confusion" between purple and blue that tied together his findings without granting the conclusion his adversary had urged: "The purple paper emits significant red and blue waves. Because the red component is not seen by the bee, only the blue element remains . . . and in this way, a purple or blue, which to us looks so very different, appears similar or identical to bees."[34]

With an explicit nod to Sprengel, von Frisch related these findings back to flowers. In areas where pollination depends on honeybees, he argued, red flowers are rare. Conversely, purple flowers more commonly grow in these regions. Bees, he explained, see the strong blue component in the purple flowers. Poppies with their bright-red petals provide a notable exception, but one that further proves the rule: the petals are so large and conspicuous that surely their size plays a role in attracting their pollinators even if their color is not perceived as such by bees. At the same time, he argued, red berries are commonly seen in these same parts. This too makes sense, considering that the propagation of their seeds depends on birds, rather than bees. Similarly, in regions where red flowers do grow, hummingbirds generally act as the flowers' pollinators. Although "one might be tempted to search for causes in the world of plants itself," he found this "difficult to believe" in light of the evidence.[35] Surely the vision of pollinators provided a more plausible explanation for the colors of flowers.

In addition to these findings, von Frisch also offered more casual observations gleaned outside his official experimental protocols. For example, when the bees were trained to yellow, they "keenly examined a yellow pencil I was using to take notes." "While

I held it in my fingers," he recalled, "they flew up and down along it and nearly touched it with their heads and also frequently settled on it." On another occasion, his brother happened to be writing letters at a nearby table, and his blue coat soon made him "the center of attention for the searching bees" that had previously been trained to blue. As any sensible person would, "he quickly took off his jacket and hung it farther away over a chair, which the animals then swarmed."[36] While not strictly part of what transpired atop the experimental table, these experimental "outtakes," nonetheless, could serve a powerful function. By giving his audiences a sense of in-the-moment happenstance, he could enlist them as "virtual witnesses," as though they too were present as bees examined pencils and swarmed nearby jackets.[37] The evidence suggested that bees could discern colors, regardless of whether they were properties of standardized, commercial papers or a jacket that happened to be nearby.

The potential impact of watching the bees fly to specific colors was not lost on von Frisch. Two years later, he performed these very experiments before live audiences at a meeting for the German Zoological Society in Freiburg. It was early June—perfect weather for bees—and von Frisch traveled to Freiburg a few days before the start of the conference to train animals on-site. Hertwig's former assistant, Franz Doflein, kept the insects for his own physiological work and agreed to let von Frisch use them for his demonstrations. This was fortunate, for the fish von Frisch had brought along from Munich had not survived their journey well enough to be used in the intended demonstrations.

Doflein, who observed von Frisch's preparatory work, later recalled, "The phenomenon" of the bees preferring the training colors over the proferred greys "was so conspicuous, that it was possible with little effort to produce a color picture on an autochrome plate."[38] The autochrome was one of the earliest color photography processes and had been patented by the Lumière brothers just ten years prior.[39] Later, during von Frisch's live demonstrations, the image was projected onto a wall, allowing even those in the audience who were sitting far back to "observe" the clustering of the color-trained bees on the colored squares. The projected color image was visually arresting, not just because it showed the bees congregating on color, but also because it presented the phenomenon via cutting-edge technology.

Over the course of the conference, von Frisch would perform the demonstrations over a dozen times for different groups. And again, as in the earlier telling of the experiment, the insects proved themselves singularly devoted to their training color: bees trained to blue reportedly stubbornly swarmed blue items of clothing worn by members of the audience.

The spectacle of the color-trained bees clearly left his audiences impressed. Doflein later recalled that the "experiments proceeded with such precision that von Frisch was able to convince the numerous zoologists and physiologists who happened to be at the conference of the correctness of his assumption that the bees must somehow possess the ability to distinguish colors irrespective of their brightness." He concluded, "We do not need to wait for a further development in the polemic between von Hess and von Frisch, but instead already now have the right to accept the results of [von Frisch], which harmonize beautifully with countless biological observations." Doflein was not alone in his assessment. Friedrich Stellwaag, a bee researcher in Erlangen also reviewed von Frisch's demonstrations in an essay, "On the Relationship between Light and Life," and concluded, "The old doctrine of the mutual relationship between flower colors and insects, which had seemed threatened by v. Hess, now has new support."[40]

But not everyone agreed with these favorable reviews. In a 1918 paper, von Hess dedicated a special section to "the demonstration of trained bees at the Freiburg Zoological Conference." He acknolwedged that "the 'training' experiments von Frisch performed . . . seemed to have made a special impression."[41] But his framing of "training" in scare quotes was telling, as he once again questioned the very foundation of von Frisch's claims. This time, he denied that the appearance of the bees on the blue square had anything to do with the conditioning phase of the experiment. Instead, he claimed that his younger colleague's work undermined the very conclusions he tried to draw from it. He argued that the bees were not at all trained to blue, and a control experiment in which bees were trained to feed on gray would have proved as much. Had such an experiment been performed, he argued, those bees would have also preferentially alighted on the blue papers provided they were equally bright.

Just a couple of weeks after von Frisch's demonstrations in Freiburg, the debate with von Hess gave way to far weightier conflict. As is well known, the Archduke Franz Ferdinand, heir to the Austro-Hungarian Empire, was fatally shot on June 28, 1914, and the events surrounding the assassination triggered what would become World War I. Von Frisch's work and lectureship at the University of Munich were interrupted by the advent of the war. He was declared unfit to serve in the military due to his poor vision, and the institute in Munich gradually emptied of its able-bodied young men in the early enthusiasm that surrounded the war. Von Frisch decided that "to stay with zoology in the now-empty institute in Munich was unsatisfying when elsewhere helping hands were so desperately needed." Hertwig agreed to release him from his duties in Munich so that he could answer his brother Otto's call to help at the Rudolfinerhaus, a hospital near Vienna. Otto von Frisch had served as a surgeon during the Balkan War of 1912 and was now in charge of the entire hospital. Its capacity had been ratcheted up from a modest one hundred beds before the war to four hundred. An additional six hundred beds were placed in nearby homes. The hospital was extremely short-staffed, as many former assistants had also left for the war.[42]

Conditions were dire as the wounded flooded the already over-extended hospital and staff. Many patients arrived after grueling travel from the eastern front with terrible wounds and festering infections. Because von Frisch had not finished his clinical training in medicine, he first served as a nonspecialist worker. His early tasks involved changing beds and dressings. But soon he was also pulled into surgical duty alongside his brother. Toward the end of

**FIGURE 2.4.** An image of Karl von Frisch during WWI pictured with one of the nurses at the Rudolfinerhaus Hospital. They are shown preparing medium for petri dishes in the bacteriological laboratory von Frisch established in the hospital basement. This is around the time he met another nurse, Margarete Mohr, who would become his wife. (Nachlaß Karl von Frisch, Bayerische Staatsbibliothek, Munich, ANA 540.)

his tenure at the hospital, he performed a number of supervised surgical procedures, including amputations above the knee.[43] These skills would prove valuable in his later work when he often performed surgical procedures on fish that needed to be kept alive for extended experiments.

At the same time, he also acquired valuable biological experience by setting up a bacteriological laboratory in an unused basement space of the hospital. This in-house facility allowed physicians to rapidly diagnose patients suspected of suffering from infectious diseases, such as dysentery, cholera, or typhoid. Critical decisions about whether to isolate or transfer a patient often depended on this knowledge, and time was of the essence.

During the preparation of a nurse's handbook on infectious diseases, von Frisch met a young nurse by the name of Margarete Mohr. She was the daughter of a Viennese book distributor and was spending her war service at the hospital. Von Frisch felt himself drawn to this serious young woman and asked her whether she might help him with a project on bacteriology he'd been developing alongside his clinical work. He needed someone to create illustrations of disease agents for the publication. Seemingly naive of the young man's intentions, Fräulein Mohr declared herself without any artistic talent whatsoever and assured him that she would not be of any use to his project. An older nurse witnessed the clumsy interaction and intervened. "For God's sake," she cried. "Stop acting like a country bumpkin and help the doctor with his project."[44]

Margarete finally agreed to help and produced the drawings for von Frisch's manual.[45] And there, bending over their microscope amid the uncertainties of war, the two fell in love. The courtship grew and Karl and Margarete soon became engaged to each other.

FIGURE 2.5. Margarete Mohr around the time she met her future husband, Karl von Frisch. (Nachlaß Karl von Frisch, Bayerische Staatsbibliothek, Munich, ANA 540.)

But in May of 1917 disaster struck again. Von Frisch's father suddenly passed away. Margarete had been able to meet the man who would have become her father-in-law, but the acquaintance was brief.[46] Despite the tragedy, the young couple decided not to delay their wedding, as the times felt increasingly uncertain. They married on July 20, 1917.

Shortly after the wedding, the couple left for their honeymoon in the Steiermark, where Margarete had spent happy holidays in previous years. But soon her groom grew restless. "To her disappointment, after only a week, I felt myself pulled with force to Brunnwinkl," von Frisch remembered. Prior to the outbreak of war, he had started "promising experiments" on the sense of smell in bees as a direct outgrowth of his work on their sense of color. He had compiled precious odor substances for the trials that had sat unused since he had traveled to Vienna to help his brother. The time away from the hospital now seemed too short to spend in idle honeymooning. The two promptly repacked their bags and traveled to Brunnwinkl. When von Frisch's older brother Otto heard of his youngest brother's sudden maneuver in the name of science, he immediately granted him another week's vacation from his hospital duties and instructed his new sister-in-law to divorce her husband if he did not take sufficient time from his work for at least two long walks.[47] It seems he was only half-joking; although papers were never filed, Margarete's birthday would, over the years, be the only day on which her husband took a break from his work.

Over the next summers, when he could get away from his hospital duties, von Frisch resumed his bee work in Brunnwinkl, especially on their sense of smell. Although there had been some disagreement about whether flower colors played a role in attracting animal pollinators, it was generally agreed that scent must function in this capacity. However, von Frisch lamented that very little work had been done to establish the specifics of this relationship. Indeed, there was considerable disagreement among scientists about whether scents attracted bees from a distance or only at close range, and whether the insects' sense of smell was acute or dull, especially compared to that of humans.[48]

Von Frisch again presented the problem in the larger context of the coevolution of flower scents and bee olfaction after Sprengel. Those plants that relied on wind pollination lacked the conspicuous, smelly flowers of the insect-pollinated plants, and other plants had evolved particular flower scents to better meet the predilections of their pollinators—the scent of rotting meat, for example, was known to be given off by plants that depend on carrion-feeding beetles and flies, he wrote. Other flowers were known to smell only for limited times of the day or night when their pollinators were most active.

In order to test whether bees could be lured by odor, von Frisch again set up a simple conditioning exercise. He had a series of square cardboard boxes made, each with a removable lid and a small round hole in front. The bees were to enter and exit the box through the hole, and its interior offered an enclosed (and therefore, more controllable) scent environment. To get the bees accustomed to entering the boxes, he placed sheets of honey-smeared paper on a table in a sheltered spot about sixty paces from the

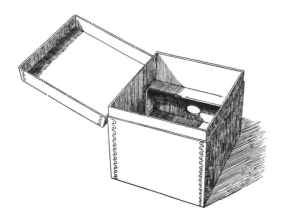

FIGURE 2.6. A drawing of one of the cardboard scent boxes von Frisch used in his experiments to test the bees' sense of smell. The cardboard strip, or "stage," could be dribbled with scented oils or sugar water to train and attract bees. The insects would enter through the round opening cut into the side of the box. (Karl von Frisch, *Aus dem Leben der Bienen* [Berlin: Springer, 1927], p. 69, fig. 45.)

hive.[49] And then he waited for the chance arrival of bees from the nearby hive. Once a single bee discovered the honey, soon others would join her. When a steady stream of ten to twenty bees at a time frequented the spot, he replaced the honey sheets with one of the cardboard boxes. He then gently coaxed the bees into the box by placing a sugar water dispenser inside, dribbling a trail of honey from the flight hole to the dispenser, and smearing its lip with honey. Once the bees consumed the honey droplets, they soon took to the sugar water.

For the subsequent, training-to-odor phase of the experiment, he replaced the experimental table with a fresh one to prevent residual honey odors from interfering with the experiment. A fresh cardboard box was also placed on the table. In addition to a clean sugar water dispenser, he also added a small cardboard ridge to the box's interior on which he applied drops of jasmine oil from a pipette. He then placed three identical empty boxes on the wooden plank next to the one containing the scent and sugar water. From time to time, he shuffled their positions on the board to prevent the bees from associating the food-containing box with a particular configuration just as he had with the test tubes and colored squares.

Once the bees established their steady search pattern, he commenced with the actual trial. He set up four fresh boxes, three of which were empty while the fourth contained a fresh cardboard strip with scented oil (but no sugar water). Once the boxes were set in place, an observer was assigned to each of the boxes. Over the course of five minutes, the visiting bees were counted and recorded. The bees were reported to exhibit clear preferences for the jasmine-scented box with 123 visits compared to only four visits to the scentless boxes. Von Frisch concluded that the bees must be able to perceive scents and, in particular, the smell of flowers.

Following further trials with bees that also included mixtures of scents, von Frisch proceeded to test himself as well as another participant whom he labeled "M.F." with the same setups he'd used with the bees. The timing and circumstances of this work make it likely that M.F. was none other than his new bride, Margarete.

FIGURE 2.7. This photograph shows von Frisch performing an odor experiment with an array of scent boxes. (Nachlaß Karl von Frisch, Bayerische Staatsbibliothek, Munich, ANA 540.)

M.F. proved to have a slightly more acute sense of smell than von Frisch. But both fell within a range comparable to that of the bees, and he concluded that the human and bee senses of smell were fairly similar, with the exception of the bees' superior performance at picking a training scent from a mixture of odors. Here the bees outperformed both von Frisch and his collaborator. "It seems, then, that the bees have to a considerable extent the ability to pick out small amounts of the training odor from a scent mixture. In this they likely surpass the average human. But it remains [an open question] whether a trained perfumer might not be capable of similar achievements."

These experiments, though generally simple in terms of their setup and the conclusions they offered, proved challenging on a practical level—odor was difficult to investigate and manage, as the threat of ambient contaminants was a constant. Even the cardboard from which the boxes were fashioned proved problematic. This became especially worrisome as he determined increasingly that the bees' ability to smell or not smell a given scent closely corresponded to that of humans. Clearly to humans, the cardboard smelled of cardboard. To address this problem, he ordered a series of ceramic boxes.[50] He gave minute specifications for how these boxes had to be constructed so that no odor clung to sharp edges, cracks, or crevices. He also offered detailed instructions on how to painstakingly clean them, and Margarete spent hours during this time carefully washing ceramic dishes and boxes in an effort to remove any residual odors. Thus, through a curious inversion of the laboratory and field, it was the "outdoors," the "free sky," and the "open air" that were believed to provide the purest and most suitable environments in which to work "cleanly" with scents.

Based on the results of this work with scent boxes and training scents, von Frisch drew two conclusions. First, the bees' sense of smell is similar to that of humans. And second, "The certainty with which the bees picked out the training scent and those scents similar to it from the large number of offered scents is remarkable and supports the conclusion that they make use of this ability to distinguish the different scents in their visits to flowers."

### Early Bee Communication Studies

In 1917, while studying the senses of color and smell in bees, von Frisch noticed something curious. Individual animals that had previously collected sugar water from a feeding dish continued to visit it intermittently even after it was empty as if to monitor its contents. When he replenished the dish's supply, within minutes a great many bees appeared at the dish. From this observation, he conjectured that the bees must somehow have alerted their hive mates when food was again available. But how? Two summers later, in 1919, he decided to pursue further the question of how bees communicate and made what he later called the "most fateful observation of [his] life": the bees' mysterious dances.[51]

That year, von Frisch had borrowed a special queen-breeding hive from the Bavarian inspector of apiculture. In such a hive, each comb was sandwiched between two glass plates that allowed observers to watch the bees' activities within. Von Frisch trained a few bees to feed at a nearby dish by luring them with honey and then marked them with red dots of paint. After keeping the dish empty for a while, he refilled it with sugar water and waited. A single bee came to the dish. After drinking, she returned to the hive, where von Frisch could observe her behavior. "I didn't believe my eyes!" he recalled.[52] The events that transpired were "so delightful and riveting" as to defy "attempts to describe them." In his first publication of the phenomenon the following year, he explained that the bee performed a kind of "dance, a 'recruitment dance,' one could call it, which brings her nearest surroundings into noticeable excitement." He continued with a detailed description of what would come to be known as the round dance: "She scurries with great speed in a circle, but at the same time often pivots by 180 degrees, so that the direction constantly changes. The circles are tight, each with one cell at its center and the bee walking on the six adjoining cells." The bee runs "one to two circles in one direction, often even just half or three-quarters of a circle, and then suddenly pivots to make turns in the opposite direction." He reported that after circling in a given location anywhere from three to thirty seconds, the bee runs to a different location where she once again begins her dance. Already in this first description, von Frisch unequivocally ascribed the function of the movements to recruitment. He noted that "the reaction it causes in the others is just as distinctive." According to his account, the bees that attended the dances immediately turned their heads toward the dancer and touched their antennae to her abdomen while following behind her. If the bee that followed was one of the marked animals—that is, one that had previously fed at the training site—then something additional happened: "Then, without concerning herself further with the [dancing bee], she rushes directly to the flight hole and to the feeding site."[53]

But in addition to observing the recruitment of bees that had previously visited the site, von Frisch made another discovery—

new animals also began to arrive at the location after attending dances. The findings prompted him to wonder whether bees led one another to the foods. In other words, after the scout bee danced, did she and her recruits fly out together or follow one another to the feeding dish? To directly monitor bees in flight posed formidable observational challenges, and von Frisch remembered having "spent much time and effort . . . to create optimal conditions for the observation." He hung white sheets in bright sunlight along the path the bees were presumed to fly from hive to food. The brightness of the sheets was to offset the insects' dark bodies so they could be seen in the air. And yet the results were negative: "I was never able to see an 'escort.'"[54]

The question remained: How exactly did recruits that had never before visited a food source find their way to it once they attended the scout bee's dances? Because of his extensive work on bee olfaction from the previous summers and his sympathies for Sprengel's theory that scent (in addition to color) served to attract bees, odor seemed a promising contender. To test its role in the bee dances, von Frisch set up experiments in which he trained bees to feed on sugar solution that rested on a surface that had been doused with strongly scented substances, such as peppermint or bergamot oil. He then set up a series of observation spots with sugar dishes. Each of the dishes rested on a base that was scented with a different oil. He found that the bees overwhelmingly favored the station that matched the scents scout bees had fed on prior to returning to the hive to dance. From this and similar observations, von Frisch concluded that the bees are alerted to the presence of profitable foods by their fellows' dances. Once they leave the hive, they search for the odor that had clung to the dancer's body. These findings fit beautifully with his holistic view of nature: "Thus, we get to know the biological significance of flower scent from another side," that is, from the perspective of bee recruitment.[55]

But von Frisch worried that the experimental setup he'd relied on in these experiments—especially the glass dishes filled with sugar water—were too contrived and therefore a poor approximation of what bees encountered in nature. In an effort to test whether scented sugar dishes provided a valid substitute for nectar, he decided to test his findings using actual flowers. Ironically, to achieve more "natural conditions," he needed greater control over the bees' feeding. He regretted that this was difficult, as "one cannot force a desired flower visit outdoors."[56] And yet just that was needed. To come closer to achieving this level of control, he relocated the entire experimental setup to a large greenhouse known as the Winter Hall at the Munich Botanic Garden. He had the space cleared of all but a few scentless green plants and then set up his experimental hive and bees. Instead of glass dishes filled with sugar solution, he now offered the bees cuttings of robinia and linden blossoms, flowers known to be rich in nectar as well as highly fragrant. He found that the bees behaved just as they had when fed with sugar solutions on scented surfaces—the returning scout bees danced, and their fellow bees that came in contact with them subsequently flew out to feed at the blooms. When he removed the flowers, the bees stopped their dances and only resumed them when they were once again offered the blossoms.

Von Frisch subsequently repeated the greenhouse experiments, but this time used poppies and roses, flowers mostly devoid of nectar but rich in pollen. When the bees returned to the experimental hive, their hind legs were packed with pollen to give the telltale appearance of "pollen shorts." Von Frisch now reported observing a different kind of dance. In a later publication, he described the phenomenon as follows: "A bee that returns home with [pollen] shorts, crawls on the comb upward and then begins to turn herself amid the other bees; but she does not describe a full circle, as is the case in the round dance, but for now only a half circle, then runs straight across 2–3 cells to her point of origin, now turns to the other side and runs a second half circle, which completes the earlier one to a full circle."[57] According to von Frisch, this second type of dance similarly served as recruitment to foods, but to pollen rather than nectar. He concluded, "It seems reasonable to see the two types of dances as different expressions of the bee language, of which the first signifies an ample nectar supply while the other means a good pollen source."[58]

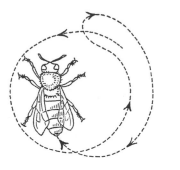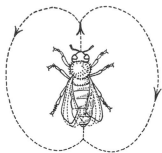

FIGURE 2.8. The circle (*left*) and waggle (*right*) dances von Frisch observed in the hive after the bees returned from foraging sites. In his pre-WWII descriptions of the movements, he believed the circle and waggle dances served to communicate nectar and pollen sources, respectively. After WWII, he would retract this interpretation and argue that the different shapes corresponded to nearby and far-away sources. (Karl von Frisch, *Bees: Their Vision, Chemical Senses, and Language* [Ithaca, NY: Cornell University Press, 1950], p. 70.)

At the time, he wished "not to overexaggerate the distinction between the two dances, for the difference between them can also get blurred."[59] And yet while he would later change his mind about the meaning of the two "different expressions of the bee language," he never abandoned the distinction between them and retained essentially a two-word conception of their communicative function. And after his initial observations in 1919 that bees seemed to alert hive mates to profitable food sources, he seems never to have asked *whether* bees communicate but only *how*. At the time, he felt the meaning of the dances to be close at hand: "I hoped to understand [it] in a few days or weeks." But instead, almost five decades of intermittent work on the bees' dance behavior followed.[60]

Before considering the consequences of this early work on von Frisch's career, it is worth dwelling for a moment on the painted dots with which he had marked his bees that summer. They were the beginning of a powerful tool that he would deploy throughout his subsequent research and that would allow him to tackle many of the issues surrounding bee work.

Bees are unlike most other animals in their need for sociality. The bee colony had long been considered a model collective with its members selflessly working toward the common good.[61] In his later book *The Life of the Honeybee*, von Frisch described the bee's dependence on its hive mates: "The farmer may keep a single cow, *one* dog, or *one* hen if he so desires, but he cannot keep a single bee—she would soon die."[62] In addition, bees within a caste resemble one another so strongly that it is virtually impossible for the unaided observer to tell them apart, much less follow their movements through the dizzying fray of the hive. But precisely this was required to investigate their dances.

To keep track of the bees, von Frisch developed a special marking system.[63] By means of a brush, paint, and steady hand, he painted little dots of different colors on the bees' thoracic and abdominal sections. The colors of the dots stood in for their numeric value while their placement on the insects' bodies indicated their decimal position. This system allowed him to distinguish and track hundreds of bees as they fed at food dishes, danced for their sisters

**FIGURE 2.9.** An early illustration of the special system von Frisch used to sequentially number his bees. Colored dots painted on the thoracic and abdominal sections of the insects allowed him to identify and track hundreds of bees at a time. (Karl von Frisch, *Über die "Sprache" der Bienen: Eine tierpsychologische Untersuchung* [Jena: G. Fischer, 1923], plate 1, fig. 4.)

and attended each other's dances, and, in turn, flew out to feed. "Because we numbered the animals, we learn something new," von Frisch stated.[64] Though simple, the technique was to become a central feature of his research and most of his work would have been impossible without it.

Von Frisch's work on fish and especially bees in the period around World War I set him on a promising trajectory. In 1921, around the time the aforementioned publications on the bee dance language appeared in print, he was offered his first professorship at the University of Rostock on the Baltic Sea. As part of his duties, he would now be in charge of his own physiological institute. The increased administrative duties alongside his full research agenda put him under far greater pressure than what he'd experienced as Hertwig's assistant. Alongside these increased claims on his time and mental capacities, his personal life had also become more demanding. While still in Munich, he and Margarete had welcomed two daughters. Johanna (known to all as Hannerl) had been born in 1918 and Maria just two years later. Von Frisch would later remember this time in Rostock with their two young daughters as one of the happiest periods of his life: "The joyous shine of our short Rostock time was never surpassed." He remembered fondly the beauty of the old city center "with its charming churches and towers." By the banks of the river, "it was picturesque . . . when the wide evening sky played in all [its] colors." He also formed lasting personal and professional ties and spent many evenings playing in a string quartet with colleagues from the university. Among the von Frisches' close friends were the ophthalmologist Hermann Pflüger and his family. Although they didn't know it at the time, the Pflügers' son, Ekkehard, would later become their son-in-law.[65]

But alongside the intellectual and personal richness they enjoyed in Rostock, private letters of the time also tell another, less joyful story. Von Frisch's wife, Margarete, struggled. Her husband's relentless schedule, their growing family, and the persistent financial worries that beset their early time together all contributed to a deep bout of depression. In May of 1922, she decided to travel to Vienna for a visit with family and to recuperate. Von Frisch wrote to his mother how he hoped the trip might both give her "the inner peace she so desperately needs," and help her realize that "she has a role to fulfill and also that she is capable of fulfilling it." He confided to his mother their private desperation: "It torments me terribly that she again and again expresses the thought that it would be better if she weren't here."[66] A year later, in June of 1923, she gave birth to a third daughter, Helene, or Leni for short.

The fall after Leni's birth, von Frisch moved his family again, this time to Breslau to assume a new professorship. He had felt himself setting roots in Rostock and regretted the move on a personal level. Nonetheless, the new position was undeniably superior. It offered a well-equipped laboratory and funds for a sizeable scientific staff. A photo of his Breslau group shows von Frisch surrounded by twenty students, collaborators, and staff members. During this time, he continued his work with the bees both at the university and especially each summer in Brunnwinkl.

FIGURE 2.10. Von Frisch (*front row, center*) shown with staff and research members of his laboratory in Breslau, circa 1924. (Nachlaß Karl von Frisch, Bayerische Staatsbibliothek, Munich, ANA 540.)

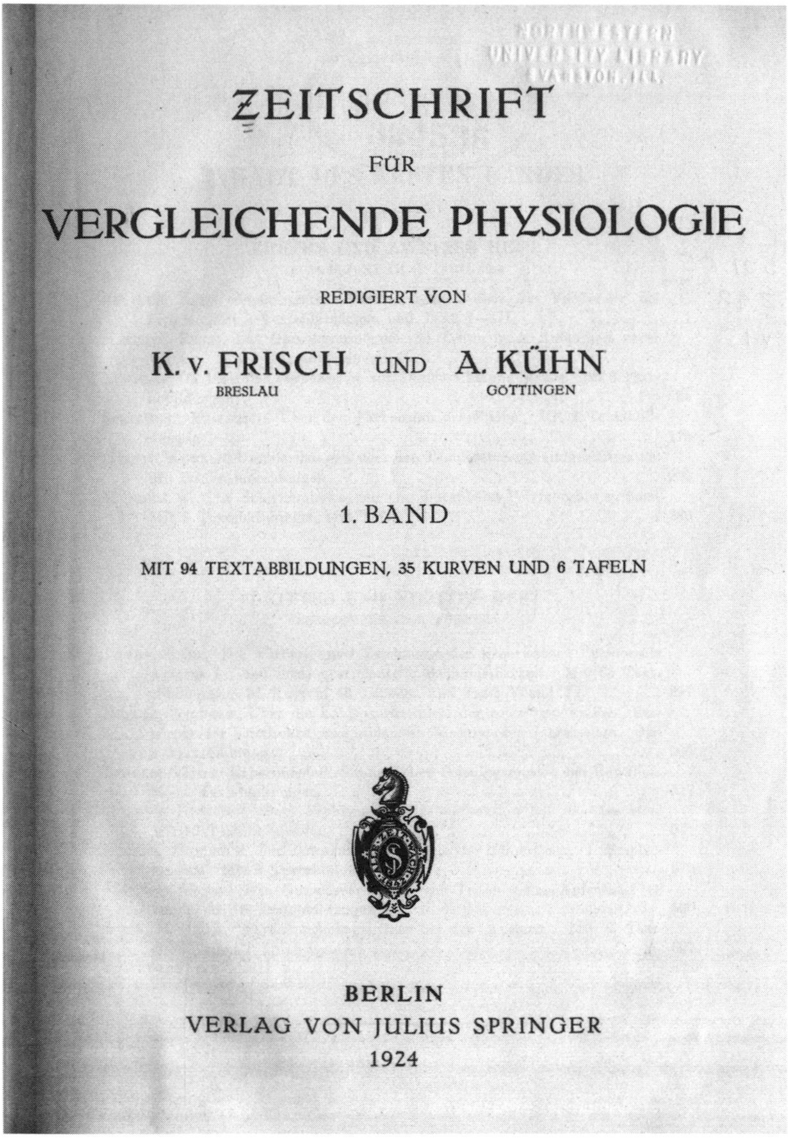

# ZEITSCHRIFT

FÜR

# VERGLEICHENDE PHYSIOLOGIE

REDIGIERT VON

## K. v. FRISCH UND A. KÜHN
BRESLAU          GÖTTINGEN

1. BAND

MIT 94 TEXTABBILDUNGEN, 35 KURVEN UND 6 TAFELN

BERLIN
VERLAG VON JULIUS SPRINGER
1924

FIGURE 2.11. The front page of the first issue of the *Zeitschrift für vergleichende Physiologie* (Journal of comparative physiology) (1924) under the editorship of von Frisch and his colleague Alfred Kühn in Göttingen.

Alongside pursuing his own research on sensory physiology, von Frisch also increasingly contributed to the larger scholarly community and dedicated himself to growing the profession of comparative physiology. In 1924, he and a colleague, Alfred Kühn at the University of Göttingen, launched the first issue of the *Zeitschrift für vergleichende Physiologie* (Journal of comparative physiology). Over the years, the journal offered an outlet for much of von Frisch's own work as well as that of his students. In addition to providing him with a publishing forum, the editorship also gave him a privileged vantage point from which to survey his nascent field. In a letter to his mother, he recounted the pleasure the job provided him: "Comparative physiology is a relatively young branch and our journal is the only one dedicated specifically to this field. Thus, the majority of what's produced in this area in Germany comes to me or my coeditor, Kühn, with whom I'm very much of a mind. I see now how this type of position, provided one takes it seriously, affords one influence over the publication practice of an entire discipline." Von Frisch clearly relished this position and took the job seriously. He and his coeditor "don't accept any rubbish," he wrote, "but we also don't reject any work without [giving] extensive objective reasons and very frequently make suggestions for changes in accepted pieces." In the days before peer review, all such work fell to editors, which must have been overwhelming at times, especially at first when the journal appeared monthly. In his usual understated tone, von Frisch conceded, "It is of course a good deal of work." But he hastened to add, "It is worth it."[67]

Today, it is easy to dismiss the dispute between von Frisch and von Hess as little more than a perturbation in the path of the former's rapidly rising star. But that would be to miss its significance at the time. Indeed, it was surely not *in spite* of his disagreement with von Hess but in large part *because* of it that he enjoyed such precipitous success in those early years of his career. The often-heated and rapid back-and-forth with the much more established scientist had a significant impact on his early publications both in quality and quantity. The persistence and scrutiny von Frisch came to expect from his adversary undoubtedly put him on his guard and made him pay close attention to his own procedures. The anticipation of a blistering attack was often visible in the very publications themselves with von Frisch explicitly addressing possible objections before they were raised. It brought a vigilance and rigor to his early work and prompted him to innovate in ways that would have a lasting effect. His introduction of training with food, standardized colors, and the aforementioned technical and visual innovations all stem from this period. It also underscored for him the importance of shoring up professional support. Hertwig's public support at this critical juncture helped the young scientist capture the attention of the larger scientific community. The early debate with von Hess at the beginning of his public research life (much like the one with a young American adversary, Adrian Wenner, that was to occur toward the end of his career) centered on clarifying the sensory capabilities of animals. Again and again, he argued that animals were capable of much more than had been previously believed. And let us not forget, von Hess's work led von Frisch to what would become his favored organism—the honeybees. While the early work with fish and bees showed great synergy between these investigations, it was his bee work that would put von Frisch on the scientific map.

After only three semesters in Breslau, von Frisch again received word of a new opportunity: he was offered the successorship of Richard Hertwig's position in Munich. Overjoyed, he accepted the position and the family moved in April of 1925. He was excited about settling back into his home away from home, and the position offered everything he could have hoped for. The university was one of the finest in Germany and the position came with generous funding for research and staff as well as the institute his mentor had led.

But just a few days after the von Frisches arrived in Munich, the family again got terrible news. His much-loved mother, Marie, had died after a brief but acute illness. Von Frisch would miss her dearly in the years to come and continued throughout his life to attribute to her much of the family's intellectual life as well as his own interest in animals.

Alongside these personal tragedies, larger storm clouds had also begun to gather over the city by the Isar. In the mid-1920s, the streets of Munich convulsed with mobs of angry men. Germany had been badly humiliated by its defeat in World War I and still bridled under the terms of the Versailles Treaty, which had brought the war to its inglorious conclusion for Germany. Unemployment climbed to vertiginous heights, and money was losing value at such a clip that images of shoppers pushing wheelbarrows filled with near-worthless notes through the streets would become emblematic of the time. Others fed paper money to their stoves in an effort to keep warm. In this atmosphere of helplessness and growing popular discontent, Adolf Hitler would soon find his political voice among Munich's malcontents.

# SENSING THE SENSES

🐝

In the 1920s and 1930s von Frisch produced a series of films intended to demonstrate the sensory abilities of animals. Each film focuses on a single animal or species and a particular sense. One proves the fish's sense of taste, another shows its ability to hear sound, and others lay bare honeybees' capacities to distinguish among colors, scents, and tastes.[1] Each film begins by showing a brief training period in which the animal undergoes Pavlovian conditioning. Once the connection between food and a conditioned stimulus has been forged, the animal is shown to react to the stimulus even in the absence of food. The films are set up much like a scientific demonstration might be, and indeed, their content parallels closely von Frisch's scientific papers from around this time.[2]

His 1927 film *Sense of Taste in Fish* opens with an intertitle:

> Training with salt,
> salt solution [is] dyed,
> to make its spread
> in the water visible[3]

The film shows a tank filled with water. A small fish swims about, visible in silhouette against a white background. The disembodied arm of a scientist with telltale white coat cuffs moves into view briefly and places a glass pipette near the top of the frame with its tip reaching into the water. A black liquid streams from the opening, snakes its way into the water, and slowly dissipates. The fish continues to swim back and forth, seemingly indifferent to the inky cloud spreading around it. A second intertitle confirms:

The fish does not react
to the salt solution;
he will now be fed with salty meat
for training [purposes]

The next scene again shows the tank, the fish, the pipette, and the scientist's arm. This time, the hand aims a glass stick at the fish. At the rod's terminus a little lump is impaled—"salty meat," we presume. Briefly we glimpse the bespectacled face that goes with the hand. The scientist peers into the tank to ensure that the fish has found its morsel. The film cuts to another intertitle:

Training success;
reaction to salt solution

Again, the tank and fish, this time in close-up. The scientist's arm reappears and secures the pipette. Black liquid streams into the water. The fish swims through the cloud, wiggles, then drops to the tank's floor, where it runs its mouth along the glass. The arm—stick and treat in hand—reappears and offers the fish its reward.

Von Frisch valued film for its power to deliver visual phenomena to live audiences and was keenly aware of the medium's rhetorical force. In 1924, he was the first to show a film of the bee dances at a meeting of the Deutsche Naturforscher und Ärzte (German Scientists and Physicians).[4] By this time, most of von Frisch's lectures to colleagues and the public included both slides and short films. Although many showed a carefully planned experiment to demonstrate a particular sense of an animal, they were not fixed products but rather constantly evolving, organic outgrowths of his work; he frequently cut and pasted existing strips to adapt to particular needs and time constraints.[5] In his memoirs, he described his first public use of the medium in 1924: "The finale was provided by a film—which at that time was still an unusual demonstration tool in scientific presentations—of the dancing bees. The spectacle of the bee dance is a fascinating event, but one cannot convey an adequate impression of it with words alone. The desire to bring the captivating image before listeners' eyes led—with the help of a skilled cinematographer—to the production of the film."[6] Thus, the power of film derived not just from its subject matter but also from the medium itself—its novelty as well as its capacity to show the motions words failed to deliver.[7]

Yet when we consider what was at stake in his films on the sensory abilities of animals, the medium becomes perhaps more problematic. We might well ask how silent, 16 mm, black-and-white film—a medium that conveys neither audio nor color, let alone odor or taste—was called upon to demonstrate the abilities of animals to distinguish among sounds, hues, smells, and tastes. Although black-and-white film registers colors as shades of gray, it is incapable of delivering the sorts of distinctions on which von Frisch's claims depended.

A closer viewing of the films' opening scenes with this question in mind reveals that during the animal training period, the audience was trained as well: invisible sensory stimuli (odor, sound, taste, or color) were represented by visual cues. Each was given its own visual signature in the black-and-white film. Film was of course not the first medium to face this synesthetic dilemma. As the late philosopher and art critic Arthur Danto has argued in his reading of Giotto's *Raising of Lazarus*, visual cues of a different register can stand in for that which cannot be shown pictorially. Unable to paint the physical smell of Lazarus's rotting body, Giotto included nose-holding women in his fresco to suggest the magnitude of both the stench and the miracle—indeed, Lazarus had been quite dead when Jesus called him back to life.[8]

In the opening of *Sense of Taste in Fish*, the audience learns to read "salty" in the blackness of the liquid spreading through the water. The intertitles tell us that the black liquid is, in fact, salty, and prepare us to distinguish between the indifferent and reacting animal. Similarly, in his film *Sense of Smell in Bees*, the audience learns that the grease stains left behind by scented oils are signs for odor. In *The Sense of Hearing in the Minnow*, each time a horn is blown, the scientist bows his head. Eventually we equate conspicuous head bowing with sound making. Finally, in *Sense of Color in Bees*, we see a blue tile stand out from a sea of gray squares because we know to associate its spatial position with its alleged color.

Thus, a second, invisible layer of meaning is revealed. While the explicit purpose of the opening scenes is to show how the animal was trained, it also serves the subtler—and arguably more important—purpose of training the viewer. With the help of intertitles, viewers' interpretations could be directed to read sounds, scents, smells, and colors in the black-and-white palette of silent film. After all, we may safely assume that the fish was already trained by the time von Frisch began filming. But if any doubts remain about the intended audience for the blackness of the salty liquid cloud, consider that von Frisch had surgically removed the fish's eyes prior to the experiment.[9]

But the training scenes serve another important function in the visual epistemology of the film. Seeing the animal ostensibly before it learned to react to food also helps establish a degree of animal agency critical to von Frisch's work. For unlike the animals of Eadweard Muybridge and Etienne-Jules Marey, which were filmed and studied solely for their mechanics of motion, von Frisch's animals were displayed to unveil the reasons behind their behavior.[10] Indeed, it is precisely because the viewer can imagine the animal not reacting—and knows what this would look like—that its reactions are meaningful. The animal's behavior in response to the stimulus is the lens through which we understand its sensory capacity. Rather than watching cells change shape under a microscope or a stylus blindly trace the physiological readings of an instrument, we see the animal's reacting body and conclude that it has, in fact, tasted a salty substance. And despite watching the animal's rather staged training, we interpret its behavior in response to this training as spontaneous. As the animal studies scholar Jonathan Burt has argued, we can ascribe to animals an inherent truth-telling imperative.[11]

Although animal agency was crucial to von Frisch's visual presentation of his experiments, it was of a constrained and tightly managed nature. For the animal's behavior to be scientifically meaningful, it had to be predictable and never random or erratic. Von Frisch's control over the animals was all the more remarkable since he never touched them directly during the demonstrations. These seemingly occult powers were especially compelling in light of the types of creatures he used—neither fish nor bees were known as trick-performing pets, nor were they common fixtures in the circus ring.

Yet the animals were not the only nonacting actors in von Frisch's films. Von Frisch himself wore the secular vestments of objectivity and truth telling—a white coat, glasses, and a serious expression. In this guise, he sagely directed the viewer's witnessing. In his 1927 film *Sense of Smell in Bees*, he opens an odorless box and holds it solemnly before the camera. It contains no bees, and no grease stains mark its interior. The message is clear: "See for yourself, the box is empty."

In von Frisch's films on sensory physiology, complex interplays between animal agency and human power, between the experimenter and the experimented, are on display. Because of the bee's or fish's status as a voluntarily behaving being, von Frisch's control is all the more compelling. And while von Frisch looks into the camera to meet and return his spectators' gaze, the animal is denied such an empathic connection. The bees' eyes are too small to be seen, and the fish, well, need we be reminded of its missing eyes?[12]

# Calm before the Storm

Over the courtyard walls of the Zoological Institute, the twin towers of Munich's Frauenkirche reached into the winter sky. It was the semester break in 1932. Deliberate scientific activity was giving way to the bustle and chaos of packing for transfer to the new laboratory. Students and assistants cleared bookshelves and packed papers into boxes. The remaining animals and instruments would have to be readied and crated soon as well.[1] As von Frisch prepared for the move, he must have paused to one more time take in the old institute. The former Jesuit monastery had offered a refuge from much of the political unrest and violence that now engulfed the city and nation. The familiar walls and rooms also housed many of von Frisch's memories.

The institute had flourished since he took over its directorship seven years earlier in 1925. The research staff, which included students, fellows, and various assistants, had grown from thirty to over sixty. Over the years, he would mentor more than one hundred graduate students. He and his students published productively and widely on the sensory physiology of animals. But as the number of students and staff grew ever larger, the venerable old building with its overgrown courtyard became less and less adequate. At times, experimental basins overflowed and flooded the floors. The water would seep through the ceiling and into the rooms below. Doctoral students decamped to the small reading desks in the library because space was in such short supply, and the hallway had been turned into a makeshift laboratory where technicians and assistants prepared countless cultures and microscope slides for the institute. In addition to these semipermanent occupants, instructors and students also ebbed and flowed into the cramped hallway to hold their introductory laboratory courses.[2]

When Augustus Trowbridge, a representative of the American Rockefeller Foundation, visited von Frisch's laboratory in 1926, he was there to see whether the foundation might fund some research and students to travel abroad for their work. The still relatively new foundation was bringing generous funding not only to US researchers but also to scientists abroad who worked on topics of public interest, such as nutrition and public health.[3] Von Frisch ushered his distinguished guest through the institute just as the laboratory course was in session in the hallway. As the two men filed past the students and equipment crammed into the tight space, von Frisch joked that the Rockefeller Foundation should just build him a new institute. Trowbridge didn't laugh. Instead, he urged von Frisch to write up a proposal and request funds for a new building from the foundation.[4]

Over the next years, intense negotiations among von Frisch, the Rockefeller Foundation, the University of Munich, and the city ensued.[5] In early spring of 1930, during the Easter holiday and just after von Frisch had submitted his final application to the foundation, he went on an extensive lecture tour of the United States to gain ideas from other institutes and deliver lectures on his work on the senses of fish and bees.

The force behind the trip had been the American-born biologist Marcella Boveri, the widow of the famous German developmental biologist Theodor Boveri. After her husband died in 1915, Marcella had returned to the United States and taken a professorship at Albertus Magnus College in New Haven, Connecticut. Even after her move, she maintained strong links to her European family members as well as former colleagues. With funding from the International Education Board and the Rockefeller Foundation, she put together a US itinerary that would take von Frisch to many of the most renowned universities in the country.[6]

The very fact that von Frisch was being considered for such high-level funding by the foundation and was being brought to the United States to speak helped consolidate his reputation at home and abroad as a first-rate scientist. In a letter to the Rockefeller Foundation in support of von Frisch's trip, the Yale biologist Ross Harrison wrote glowingly that von Frisch's "qualification and standing must be well known to the International Board from the consideration that has been given to the project of building a new Zoological Institute in Munich." He continued: "I need, therefore, only say here that he is one of the leading zoologists of Germany of the younger generation and has one of the most active laboratories in Europe; that his general interest is in the field of comparative physiology and his most outstanding work is on the behavior and instincts of bees; and that he has a most attractive and interesting personality." Harrison also informed the foundation that initial queries to potential hosts for the proposed trip had met with warm reception: "I am assured that he will be cordially received and invited to lecture at Harvard, Brown, Yale, Columbia, Princeton, Pennsylvania, Johns Hopkins, Cornell, Chicago and a number of other places."[7]

And so it came that over the Easter holiday of 1930, von Frisch set out on his first transatlantic voyage. His youngest child, a boy named Otto after his uncle, was just three months old and in shaky health. Margarete stayed behind to take care of him as well as their three older children. On the way, von Frisch's ship was caught in a storm. "House-high swells" tossed the vessel about and added two more days to the trip for a total of twelve. Finally, on March 10, the boat arrived safely in New York Harbor. Von Frisch was in awe: "The highest parts of the skyscrapers pierced the ground fog that enveloped the city," and winds "tossed shreds of fog" about "the upper stories." The spectacle reminded him of the mountaintops of the Alps.[8]

After disembarking in New York, von Frisch traveled to Yale University in New Haven, Connecticut, to meet with Marcella Boveri and her Yale colleague Ross Harrison. There, amid a small group of sympathetic listeners, von Frisch gave his first professional talk in English. Initially, he clung to his notes. But soon he found his footing and once again spoke freely without having to refer to the written script.

After Yale, von Frisch traveled first to Harvard, then to Columbia in New York City. Next, the itinerary took him upstate to Cornell and then to Buffalo with a side trip to Niagara Falls. The journey then proceeded into the heart of the Midwest to Ann Arbor and then Chicago. While in Chicago, he "witnessed zoology revealed from a different perspective" when he visited one of the city's famous meatpacking houses.[9] Almost three decades after Upton Sinclair had described how millions of cattle were driven from a "sea of pens" along a "very river of death" with terrifying efficiency, von Frisch observed the assembly line that "transformed the living pig into the ready-to-ship can."[10] He astutely noted how Chicago held the lucrative role of gateway between the "overproduction of cattle of the western agrarian states" and the "eastern centers of consumption." To his European sensibilities, the site exemplified a uniquely "American genius for business."[11]

After Chicago, von Frisch traveled to Madison and then to Iowa and Minneapolis to speak at their respective universities. In Bloomington, Indiana, the famous sex researcher Alfred Kinsey hosted him and took him spelunking along "subterranean rivers with their strange animal life."[12] Next, he traveled to Columbus and Cleveland, and then back to the East Coast, with stops in Philadelphia, Princeton, and Washington, DC.

Altogether, he visited nineteen colleges and universities as well as the Bureau of Entomology in Washington, DC, the Rockefeller Institute, and the Woods Hole Marine Biological Station. Von Frisch reported to the Foreign Office (Auswärtiges Amt) in Berlin, "At all the universities, I thoroughly inspected the biological institutes and often also had the opportunity to visit the related fields (psychology, botany, physiology, etc.)." His assessment of the quality of the facilities he visited was somewhat mixed. While he considered many of the zoological institutes "*below* the average level of our German institutes in terms of equipment and scientific operations," he also acknowledged the superiority of others: "I encountered excellently equipped facilities with extraordinarily lively and diverse scientific operations unlike what we have in Germany." He marveled how "everywhere the desire to modernize the institutes is unmistakable, and significant resources are being dedicated, not only to expanding the facilities but also to equipping [them] with instruments and hiring a tremendous [number of] scientific staff." He felt the writing on the wall for German science: "It is certain that the United States will take the lead and leave us behind in the foreseeable future in ours as well as other subjects, if there are not also energetic steps undertaken in Germany to expand and modernize scientific institutes."[13]

Overall, von Frisch was pleased with his trip and felt that the "main purpose of my visit, to inspect the American institutes for inspiration for the design of our new Zoological Institute, was satisfied to the fullest degree." He'd been received "everywhere with the greatest hospitality and kindness." He was also gratified with how his lectures had been received, and noted in a letter to his wife that audiences had spontaneously started to clap at certain moments during the talks.[14] He also considered "the number of new and potentially fruitful personal contacts I was able to make in the short time of my stay a special benefit of this trip."[15]

But the highlight of his trip was an unexpected appearance by Augustus Trowbridge of the Rockefeller Foundation at a talk he gave in Princeton, New Jersey. Trowbridge had decided to deliver the good news in person: the foundation had approved von Frisch's proposal to fund the construction of a new institute of zoology. Altogether, 1,553,000 reichsmark (approximately $372,000 at the time) would be paid to the University of Munich over the next three years.[16] Of this amount, 993,000 reichsmark would go directly to von Frisch for the construction and running of the new institute. The remainder would fund a new institute of physical chemistry. The state of Bavaria, for its part, would donate the centrally located land on which the new institutes were to be built, a pledge valued at 1,050,950 reichsmark.

Von Frisch was elated and returned home in late April ready to begin planning the new facility. Over the next two years, he worked closely with an architect to design a space that would unite both form and function for the purpose of science. Set on three levels and filled with state-of-the-art equipment, the Rockefeller-funded building was a marvel to behold—it boasted separate rooms for tissue culture, large aquaria in special water rooms, and separate, climate-controlled chambers that ensured animals could be bred and kept under constant temperature and moisture conditions.[17] Other rooms housed the institute library as well as skeletons and prepared specimens from the teaching collection. Throughout the institute, what must have seemed like miles of lab benches awaited the scientists who would soon move in to pursue their craft. On the second floor, a tea kitchen was installed for social gatherings and informal discussions. The building was complete in 1933. If the spirit of science had built itself an ideal home, then surely this would have been it.

But during the move into the new quarters, ominous events loomed over the otherwise happy occasion. During the summer of 1932, the Nazis had doubled their seats in the Reichstag and their leader, Adolf Hitler, was making noises for the chancellorship. On January 30, 1933, his bid was successful and he was named chancellor of Germany.

The change in government almost immediately affected the university.[18] The Nazis cared about education because youth were seen as the key to continued power in the Thousand Year Reich. The University of Munich had long been a site for anti-Semitic agitation even before it became official government policy through Hitler's seizure of power.[19] Students were among Hitler's early support base, as many were right-wing and profoundly nationalist. With so many Germans unemployed, young men at universities counted themselves among the lucky. But Nazi student organizations played to their fears about the future and stoked their anger against Bolsheviks and Jews as well as the perceived injustices of Versailles. They harassed professors during lectures and spread nationalistic and anti-Semitic leaflets throughout campus to recruit members and remind others of their mission and ominous presence.

In contrast to the students, the professorate was considered largely Jewish, although actual numbers varied greatly by field and were lower than the rhetoric of the time suggested (33 out of 384 instructors in Munich in 1933, roughly 8.6 percent). The faculty had previously governed itself; that is, departments and deans de-

FIGURE 3.1. The library and reading room housed the new Zoological Institute's journals and books. Other rooms contained taxidermic and microscopic slide collections. (Karl von Frisch and Theodor Kollmann, *Der Neubau des Zoologischen Instituts der Universität München* [Munich: A. Huber, 1935], p. 17, fig. 17.)

cided matters of hiring, firing, and promotion. But with their rise to power, the Nazis increasingly restructured the university after the *Führerprinzip*. This meant that the president of the university officially served as its *Führer*, or absolute leader. Nazi officials, in turn, selected the president from a pool of candidates they deemed politically reliable.[20]

New laws designed to purge and Aryanize the university further determined decisions over university personnel. In April of 1933, just two months after Hitler took power, the Nazis passed the Law for the Restoration of the Professional Civil Service.[21] In an effort to establish a loyal and compliant civil service, the law aimed to rid its ranks of any members who were deemed politically or racially questionable. Jews and others who were considered undesirable by the Nazis could be legally ousted from their government and university positions. The third paragraph—which came to be known as the Aryan Paragraph—required all civil servants to at-

FIGURE 3.2. The new institute's aquarium room with tanks and workbenches. The room was part of an extensive salt- and freshwater aquarium system with indoor and outdoor basins that allowed for the study of aquatic organisms under different light, temperature, and water conditions. (Karl von Frisch and Theodor Kollmann, *Der Neubau des Zoologischen Instituts der Universität München* [Munich: A. Huber, 1935], p. 21, fig. 24.)

test to their Aryan descent. In addition to filling out a form about their religious backgrounds, the law required them to supply their own birth certificate as well as the birth and baptismal records of parents and grandparents. The law made a few exceptions—those whose appointments had preceded the outbreak of WWI or who had fought at the front could remain in their posts even if they were non-Aryan. But even these reprieves would not last long.

When von Frisch filled out the required questionnaire in 1933, he listed himself, as well as all members of his family, as Catholic.[22] He dutifully returned the forms to the university by mid-June 1933. For the time being, it seemed, his position within the university was secure.

Others were not so lucky. At the University of Munich alone, the initial purge of 1933 resulted in the dismissal of twenty-four instructors (roughly 6.3 percent of the faculty).[23] More would follow over the years. These dismissals of Jews and "enemies of the state" resulted in vacant positions that could now be filled with individuals more palatable to the regime. Von Frisch's institute was assigned such "hires," and members of his staff remembered no longer feeling safe to speak freely. Certain persons, they suspected, had been charged with listening to careless conversations and with reporting these to the authorities.[24]

Although denunciations often took the form of vague rumors and hearsay, it was never clear what would stick, and run-ins with officials could bring anyone to the brink of peril. Indeed, denunciations were so numerous among all sectors of society that even officials responsible for hearing such cases spoke of an "ugly denunciation fervor" fueled by "purely personal vendettas."[25]

Students, too, often posed a threat to professors they considered politically weak or questionable. Shortly after the Nazis took control, von Frisch was summoned to the Ministry of Education and Culture for questioning. He learned that some of the students in his laboratory course had filed complaints against him. He had required them to perform an earthworm dissection, and although the worms had been anaesthetized, some of them moved when they were cut. Now he stood accused of cruelty to animals. In a perverse twist of morals, the Nazis cared emphatically about the rights of animals and almost immediately after the seizure of power introduced legislation to protect them.[26] When questioned about the alleged abuse, von Frisch inquired why the ministry was not taking action against fishermen who impaled their worms without the use of anesthetics. Those worms, the official informed him, were being sacrificed for food and therefore died in the service of the Reich.[27]

Von Frisch was released from the conversation without any serious consequences. But the incident made clear that the tables had turned—nobody was safe from denunciation and even once-powerful professors had to tread carefully so as not to offend.

In early December of 1934, the students of Munich struck again. An anonymous poem entitled "The Neutral Scholar" ran in a Nazi student paper.[28] In it, a professor who studied bees was lambasted for pontificating readily and at length on the insect state. But when this same professor was asked after his own, human *Volk*, he suddenly fell "silent as a mouse." The piece went on to accuse the scholar of only caring about his narrow specialty and preserving the "freedom of science." "Why would he serve his own Volk," it demanded pointedly, when "the bees, bugs, lice, and fleas" were far more important to him? And "such a man wears the PhD's cap," it declared with disgust. The moral of the poem was made explicit in the closing lines: "The German demands our respect and admiration" much more so than the "doctor and scholar." While the piece did not explicitly name von Frisch, its disdainful rhyming lines left little question about their intended target.

By 1936, von Frisch's university file also contained two incriminating pieces of information. In the first, a Dr. Karl Eller recalled in a statement that he had gone to school with von Frisch's niece Brigitte Exner von Erwarten, and she had confided in him that she had "Jewish blood in her veins." Therefore, he reasoned, the same must be true of her uncle Karl von Frisch. Another statement, by a Dr. Scharnke, informed that his colleague in the storm troopers, Bernhard Krauss, had repeatedly assured him that at least one of von Frisch's grandmothers was Jewish.[29] Whether these rumors in von Frisch's file were true was of little consequence. What mattered was that these statements had been submitted together with a request dated April 7, 1936, by Wilhelm Führer, who led the Instructors League and was a fervent anti-Semite.

The Instructors League was a National Socialist organization with branches at each of the universities. The Munich league, headed by Führer, was especially active. Its numbers drew mostly from younger members of the faculty, many of whom did not yet have permanent university positions. The Instructors League took an interest, not just in these assistant professors, but in all personnel matters within the university. In Munich, the league became a powerful clearinghouse on questions of political reliability and character, and the president and deans often conferred with them on personnel decisions and would generally defer to their judgment.[30]

Führer's letter demanded that the University of Munich president, Leopold Kölbl, initiate an investigation into Karl von Frisch's possible non-Aryan bloodlines. He claimed that von Frisch was known to display a "downright small-minded opposition to anti-Semitism" and offered "unusually great partiality for Jews and Jewish associates by marriage" in his laboratory. In addition to these shortcomings in attitude and comportment, he also reported the more damning claim that von Frisch was "generally considered of Jewish blood." He was at least one-quarter Jewish according to a relative "who knows the family history precisely."[31] In response to Führer's agitations, Kölbl submitted a request to the State Ministry of Education and Culture that it contact the Reich Genealogical Department and begin inquiries into von Frisch's heritage.[32]

By 1937, the rumors and allegations had begun to affect von Frisch's public profile. In that year, he agreed to give a series of public lectures on his work on animal senses for the Nazi cultural agency, Amt Rosenberg. As part of the National Socialist politically inflected cultural programming, the Amt Rosenberg established a range of public offerings for domestic and international audiences. Germany boasted a well-established tradition of scientists writing and speaking about their scholarly work to lay audiences.[33] But in the prevailing disconnect between the Nazi commitments to appearing cultured and on the forefront of science, on the one hand, and their pronounced hostility to free thinking, on the other, officials closely monitored speakers and carefully vetted their talks to ensure compliance.

Von Frisch undoubtedly thought of the content of his proposed talks on the sensory physiology of animals as thoroughly apolitical. Indeed, he considered science, by its very nature, to be above the fray of politics and personal interests, a view that in retrospect appears either willfully naive or at the very least tinged by a strong dose of wishful thinking. It is also possible that he felt this public service for the Nazi cultural office would reflect favorably on him. Whatever von Frisch's expectations, the Amt Rosenberg now rejected his application.

One of the reviewers declared von Frisch incapable of presenting his ideas in a way that made sense to the public, and claimed that his research was of the driest, most esoteric sort sure to do more harm than good to efforts aimed at recruiting the lay public to the splendors of German science.[34] The characterization of von Frisch's shortcomings smacks of National Socialists' descriptions of what they considered "Jewish" science, exemplified most famously by how Albert Einstein's work was criticized for being overly theoretical and abstract. In addition, the accusation that von Frisch's work was inaccessible to lay audiences would have seemed preposterous to anybody who had ever heard him speak and was incongruous with his high demand as a lecturer to wide-ranging audiences.

In addition to his public lecturers, von Frisch also wrote several books and articles for lay audiences and edited a series of small books under the title "Understandable Science" that specifically aimed to bring the work of well-known scientists to the public. His own contribution to the collection, *The Dancing Bees*, was already in its second edition by this time and would go through eight more over the coming years.[35] In 1936, von Frisch had also published a biology textbook called *You and Life: A Biology for Everyone*, which was underwritten by Propaganda Minister Joseph Goebbels's special fund to create reading materials for the troops.[36] Thus, von Frisch's credentials for conveying scientific knowledge in a way that was both clear and interesting to lay audiences surely were sound.

The real reasons he was deemed unfit for public speaking on behalf of the Amt Rosenberg were, of course, political. In response to queries the Amt had sent out to various offices about von Frisch's background, officials at the University of Munich reported back that he was on their books as one-eighth Jewish but that he was most likely one-quarter Jewish. The Amt Rosenberg, in turn, now passed along this information from the university to those whom they had queried earlier who had failed to raise objections to von Frisch's candidacy as a speaker. Curiously, the Instructors League had reported to the Amt that they knew of no cause for concern about von Frisch. It is unclear why the league did not indicate their earlier questions about his ancestry, but it seems likely that they had not turned up any further evidence against him.

But now, the Amt Rosenberg informed them of von Frisch's suspected Jewish ancestry and added that he used to say in his lectures, "A shot of Jewish blood could only help improve the German race, which is too languid as it is." Von Frisch, they claimed, made this statement in reference to his own "Jewish blood admixture."[37] Whether von Frisch ever made such statements is unknown. But judging from his writings and personal letters, it seems unlikely that he would have risked antagonizing his students in such a public manner. Yet such claims appeared again and again in his file alongside what were supposed to be the "facts" of his ancestry in an ugly mix of rumormongering and racially motivated bureaucracy. The incident is a perfect example of how multiple loops of information and misinformation could swirl around individuals and gain dangerous momentum through hearsay and repetition.

The intensification of von Frisch's scrutiny by both the university and political offices was also linked to the 1937 passage of the German Civil Servant Law, an amendment to the 1933 Law for the Restoration of the Professional Civil Service. The opening paragraph called for "a professional civil service rooted in the German Volk" and "penetrated by the National Socialist *Weltanschauung*, which is bound in loyalty to Adolf Hitler the Führer of the German Reich and Volk." It demanded "true love of fatherland, readiness for sacrifice, and complete dedication of labor" and required all civil servants to swear an oath of allegiance to the Führer.[38] Paragraph 72 of the law ruled that those who were non-Aryan or married to someone of non-Aryan descent would be retired, provided they were unaware of their non-Aryan status. If they had knowingly withheld information about their non-Aryan roots, they would be immediately dismissed.[39] Once the law went into effect, all university professors were again required to provide proof of their Aryan ancestry.

This time, von Frisch attached a statement to his completed questionnaire that explained that his maternal grandmother's background was uncertain. In retrospect, it is possible that this drew attention to an already delicate situation. Certainly his attempt to speak on behalf of the Amt Rosenberg ensured that the information about his ancestry would be more widely circulated. And indeed, after 1937, von Frisch received repeated requests to furnish records for his maternal great-grandparents. But though they searched far and wide, neither von Frisch nor his brother Otto were able to locate the documents.

Remarkably, at the same time that these political agitations were underway against von Frisch, he and his lab continued their productive research program on the sensory physiology of animals. In 1937, just as the situation was heating up in the search for documents about his ancestry, von Frisch published a paper in the inaugural issue of the *Zeitschrift für Tierpsychologie* (Journal of animal psychology).[40] The new journal was to serve as an outlet for animal psychology and included among its editors von Frisch's old friend and former schoolmate Otto Koehler as well as the up-and-coming animal behaviorist Konrad Lorenz. In the paper, von Frisch identified the most important work he and his collaborators had performed over the previous decade with a particular emphasis on the relevance it might bear to human psychology. The overwhelming majority of the cited works had been published in his own journal, *Zeitschrift für vergleichende Physiologie* (Journal of comparative physiology).[41]

Like many of von Frisch's early writings on bees, the paper began with a discussion of the makeup of the colony. As had been observed for centuries, the hive is governed by a strict division of labor, especially within the ranks of female worker bees. Some clean the hive, others take care of the brood, and still others fly out to collect nectar and honey. Although it was not clear how these tasks were assigned or chosen among the bees, it was thought that each of the workers rigidly adhered to her specific task.[42]

But von Frisch's former graduate student and later assistant Gustav Rösch had investigated these phenomena and determined that each worker bee passes through a series of stages during which she performs all the different tasks of the hive, from cleaning to tending larvae to foraging for pollen and nectar. And through the course of her development, a bee's physiology changes in keeping with the tasks she performs. Here von Frisch emphasized how this arrangement differed from human society, where members choose a profession and then tend to stay with it through the course of their lives. The bees' "division of labor," by contrast "is fundamentally different from human practice. There are no master builders and brood nurses, guards and food gatherers, who practice their occupation for the duration of their lives. Rather, each worker bee performs several occupations in law-governed, temporal succession according to a certain dependence on her changing bodily development."[43]

The question then for von Frisch, and indeed the one he felt psychologists would be most interested in, was whether this transition of duties was determined by the animals changing physiology or whether the physiology changed in response to the animals' tasks. In the second case, the animal should have the ability to adapt its function at least somewhat to the needs of the colony. Gustav Rösch had investigated precisely this question through an ingenious method that effectively allowed him to split a hive into younger and older worker bees. He found that there was a certain amount of flexibility within the colony that allowed bees to respond to changing needs by reversing some of the physiological changes that had occurred as the animals had progressed to later tasks. When von Frisch recounted these results, he acknowledged that "one is tempted to say: the will governs the body." But despite the temptation, he characteristically urged restraint: "We know nothing of the will of the bee and leave the mystery unsolved."[44]

Von Frisch went on to review experiments about the bees' ability to distinguish shapes by several researchers, but especially by one active in Berlin in Richard Goldschmidt's Kaiser Wilhelm Institute. Her name was Mathilde Hertz, and she had been one of the last students to complete her dissertation in zoology with von Frisch's former mentor Richard Hertwig. Later she would become one of the few women to complete a *Habilitation* (postdoctoral thesis) in Berlin. Her scientific fame preceded her—her late father had been the famous physicist Heinrich Hertz, who had discovered electromagnetic waves (and after whom the unit of frequency, the Hertz, was named).

Mathilde Hertz was working in Munich when von Frisch first

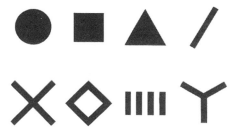

FIGURE 3.3. Figures used by Mathilde Hertz in her experiments with bees' abilities to recognize shapes. She argued that the insects could differentiate between shapes belonging to the top and bottom rows but not among the shapes within a given row. (Karl von Frisch, *Bees: Their Vision, Chemical Senses, and Language* [Ithaca, NY: Cornell University Press, 1950], p. 23, fig. 14.)

took over the leadership of the institute. She was a talented researcher who was especially respected by the Gestalt psychologists of the period.[45] Von Frisch was particularly interested in her work on bees' abilities to recognize shapes, colors, and contrasting colors, and her experiments based on shape recognition and food training had revealed surprising results. She argued that although bees were capable of distinguishing between some shapes, such as a solidly colored circle and a crisscross of lines, they seemed unable to tell that same circle from a solid triangle or square. By the same token, a Y-shaped figure was indistinguishable from a set of four parallel lines. Hertz came up with a number of shapes and grouped them into two categories.

She concluded somewhat vaguely that bees must perceive figures according to the degree of complexity of their structure (*Gliederung*). Thus the insects would distinguish between strongly structured, that is, "contour-rich," structures and solid and relatively "contour-poor" structures. The experiment pointed to differences between what humans and bees find "similar" or the "same" while also suggesting that there might be problems in human interpretations of experiments on animals. Von Frisch recognized as much: "This leads to the bees not being able to solve many problems of form perception, which are easy by human standards." The vagueness of the term *Gliederung* also suggests the difficulties in negotiating such differences between the systems of apperception of humans and animals. Von Frisch mused on this difference: "Those who concern themselves only with the bees' apperception of forms, might think that a comparison between theirs and the capacities of human sensory organs meets with large difficulties."[46]

Von Frisch's objection was curiously weak—not much more than an oblique swipe at an explanation that offended his sensibilities more than it contradicted observations. As always, the notion of a simple compulsion in animals was too blunt an explanation for his liking and oversimplified what he believed to be very complex processes. He preferred to give an explanation of the bees' behaviors that took into account their evolutionary development: "On the whole, one has the impression that the bee species spontaneously prefers or more easily learns those shapes which are familiar to it in nature and which it has assimilated into its wealth of experience through countless generations." "Physical shapes," he continued, "are far superior to flat surfaces; small black dots, which remind of the apexes of corn stalks, attract more than do large black spots; flower-like figures more than flower-unlike ones."[47]

In the paper's final section, von Frisch discussed what he considered "perhaps the most amazing result of all the bee studies of the past years"—the bees' internal clock. In 1929, one of von Frisch's students, Inge Beling, discovered that when bees fed at a particular time of day for several days, they would soon begin to appear at the feeding site even slightly before that time on subsequent days. Subsequent experiments to determine whether their internal clock derived from external factors—such as the sun's position with respect to the horizon—or some internal rhythm found the latter to be the case. Even bees that were kept and fed in a cave far away from natural sunlight consistently maintained a 24-hour rhythm. Von Frisch wondered: "Is this capacity of the bees a senseless gift of nature or does their internal clock have a biological significance?"[48] In the end, he decided that the behavior must have a purpose, which was in keeping with his vision of purposive nature.

Overall, the publication was clearly aimed at a more psychologically oriented readership than the one served by the *Zeitschrift für vergleichende Physiologie* under his coeditorship. Von Frisch believed that the most useful lessons for psychologists would be the areas of contact, where bees behaved the same or differently from humans based on the insects' specific sensory physiology: "Of the rich harvest of results, those which point to the fundamental accord or discord with human sensory physiology are of special interest to the psychologist." He added: "We have attempted to venture from the solid ground of sensory physiology onto psychological ground. We are pleased [about] where [we] succeed in uncovering new connections and in popularizing the charming life in the bee hive." But at the same time, he rejected speculations about the inner workings of animals because he believed that mental processes, human or nonhuman, were not accessible to rigorous scientific probing. Ultimately, this inability rested on physiological grounds. We might infer what goes on in other humans because we share a largely similar physiological makeup. With other animals, however, we cannot surmise what goes on in their minds because different species experience the world in fundamentally different ways. Despite having revealed interesting points of departure from physiology to psychology, "we also mind the veil behind which the bees' 'soul' will always remain hidden."[49]

What is perhaps most striking in this paper, especially in light of the events that were about to transpire, was the complete absence of any reference to the practical applications of this work. The problems presented here were clearly ones von Frisch and his students pursued out of genuine intellectual interest. But this was soon to change, as members of his lab were increasingly affected by the politics of the time.

In January of 1933, von Frisch had written to a promising young scientist, Curt Stern. At the time, Stern was visiting Caltech in Pasadena, California, from the prestigious Kaiser Wilhelm Institute in Berlin, where he worked in the geneticist Richard Goldschmidt's section. In his letter, von Frisch offered him the equivalent of a tenured position and stressed the benefits of the job as well as the important expertise in genetics Stern could bring to the Munich lab.[50] But the appointment fell through; Stern was blocked by the Nazi Ministry of Education because he was Jewish. Unlike many others, Stern grasped the gravity of the situation quickly and resigned his position from the Kaiser Wilhelm Institute. In a letter that was widely circulated among biologists, he expressed the capricious unfairness of the situation: "The new law excludes from scientific work not particular persons who have demonstrated that they are unworthy of participating, but all 'non-Aryans' without regard for their accomplishments and efforts."[51] Stern immigrated to the United States, where he would become professor at the University of Rochester and later at the University of California, Berkeley. Goldschmidt himself would leave the Kaiser Wilhelm Institute just two years later in 1935. Von Frisch recalled how Goldschmidt had informed him of his plans to leave deep in the basement of the new Zoological Institute, where they could be sure that nobody would overhear them.[52] Under growing pressure from the regime, von Frisch's Jewish assistant, Dora Ilse, also left, moving to England in 1936.[53] She would remain professionally close with von Frisch and translated his *Dancing Bees* into English for the British publisher Methuen after the war. Von Frisch's colleague Mathilde Hertz, who had researched the bees' ability to distinguish shape and color, was also forced to leave her position at the Kaiser Wilhelm Institute for Biology and the University of Berlin; she immigrated to Cambridge, England, in 1936, where she would lead an increasingly isolated and impoverished existence.[54]

In the beginning of 1940, von Frisch received another plea for help. A letter from a Dr. Halina Wojtusiakova, the wife of Roman Wojtusiak, a Polish biologist and professor at Jagiellonian University in Krakow, asked him to intervene in her husband's recent arrest.[55] The Nazis had taken Wojtusiak prisoner at the university in November of 1939, shortly after they had invaded and annexed their eastern neighbor. The mass arrest and subsequent deportation of some 183 members of the university was part of the German plan to "liquidate" the Polish intelligentsia. Roman Wojtusiak and his fellow captives from the university were transported to the concentration camps Sachsenhausen and later Dauchau, near Munich. Von Frisch wrote a postcard to Wojtusiak's anxious wife in April of 1940: "I tried various things, but only a few days ago did I probably reach the right place and speak with a man who is able to talk on-site with the appropriate persons." In typical von Frisch fashion, he closed his missive reassuringly and in the name of science: "If there is no concrete evidence against him, then I believe that he will soon again be able to dedicate himself to his scientific works."

Wojtusiak had spent a postdoctoral year in 1932 in the labs of von Frisch and his coeditor of the *Zeitschrift für vergleichende Physiologie*, Alfred Kühn, to whom Halina had also written after her husband's arrest. During his time with Kühn in Göttingen, Wojtusiak had made the fleeting acquaintance of Walter Greite, a student in the final stages of his doctorate. After the Nazis came to power, Greite climbed to officer rank in the SS and became a senior civil servant in the Nazi Ahnenerbe under Heinrich Himmler. He now visited Dachau in spring of 1940 to try to effect Wojtusiak's release. Greite explained during a brief meeting in the concentration camp that he had come from Munich "at the behest of Professor Karl von Frisch."

Although the exact details are unknown, Wojtusiak was released from the concentration camp on September 8, 1940. Others had also spoken out on behalf of the Polish intellectuals and the international press had reported the arrests and deportations with outrage. By January of 1941, the other inmates from the Krakow university were released as well, although ten of the original 183 captives had already died or would soon after their release as a consequence of the harsh conditions of their captivity. The researchers who carefully unearthed this connection between von Frisch and Wojtusiak noted that never again would such a large group of concentration camp inmates be freed: "[The release] would remain a unique event in the terrible history of the German concentration and extermination camps."

But just around the time that the final inmates from the Polish university were released, von Frisch's own life would take a turn for the worse. He received a letter in early January 1941 from the president of the University of Munich. The letter communicated a statement that had been sent to the president by the Bavarian Ministry of Education and Culture. The note informed that "the full (*ordentliche*) Professor at the University of Munich, Dr. Karl von Frisch, in according to the findings of the Reich Ministry of Education is a second-degree crossbreed. The Minister for Science, Education, and Public Instruction intends therefore to retire him according to paragraph 72 of the DBG [German Civil Servant Law]. I request that Professor Dr. Karl von Frisch be informed of this intention."[56]

# In the Service of the Reich

From the very beginning of his ordeal, von Frisch considered the challenge to his job as much a test of nerves as one of public relations. He wrote to Otto Koehler: "It seems very doubtful to me that initiated efforts or petitions promise any kind of success if it is not possible to get very decisive people on board."[1] Von Frisch was well connected, but the challenge of navigating the inscrutable bureaucracy of the various ministries and offices was formidable, and the delicacy of the situation required careful weighing of each step. The key was to gather the support of as many people of influence as possible but without making the situation generally known so as not to provide additional momentum to detractors. In his letters to Koehler, von Frisch worried that if too much attention was drawn to the situation, officials in the Bavarian Ministry of Education and Culture might see themselves obliged to pursue his forced retirement even more aggressively to satisfy party members.

In addition to Koehler, von Frisch contacted three other scientists within days of receiving the news of his imminent ouster: Alfred Kühn, Fritz von Wettstein, and Hans Spemann. By now Kühn had served with von Frisch as coeditor of the *Zeitschrift für vergleichende Physiologie* (Journal of comparative physiology) for nearly two decades. He and von Wettstein were two of the most powerful men in German science; each directed a section of the Kaiser Wilhelm Institute in Berlin, and they were among the best-funded biologists with important connections to the German Emergency Society.[2] They would do all that was in their power to effect a reversal of von Frisch's fate: they wrote letters, spoke to ministers, arranged meetings, sent telegrams, strategized, made phone calls, and discretely rallied others to his cause. At the same time, they kept up a frequent correspondence in which they vacillated between urging von Frisch not to lose faith and gently encouraging him to prepare for the event that he would lose his job.

Hans Spemann also presented an august figure in German science. He had received the Nobel Prize for his work in developmental biology five years prior. Although he was of an older generation than von Frisch, he had followed the younger zoologist's career over the years with warm encouragement. In 1921, when von Frisch received the professorship in Rostock, Spemann, who had held the same post years before, had written to von Frisch, "It makes me extremely happy to hear that you received the position in Rostock." He recounted the old institute wistfully, "When I think back to my office with its view onto the old trees of the Blücherplatz and onto the St. Jacob's Church where the white seagulls fly as harbingers of the nearby sea, I could envy you." He asked von Frisch to give his "greetings to the trusty old institute."[3] Von Frisch now turned to him exactly two decades later to ask for help.

Spemann immediately replied with shock and sadness at von Frisch's news and expressed his sympathies on an even more intimate level: "That people like your mother are no longer allowed to live among us is enough to despair." Von Frisch's mother was already dead by the time the Nazis took power, but Spemann's point was clear—he considered the regime's actions beyond the pale of human decency. He urged von Frisch to keep his chin up and "not let in this poison of the times."[4]

Spemann's campaign of support included a letter to Bernhard Rust, the head of the Ministry of Science, Education and Public Instruction (Ministerium für Wissenschaft, Erziehung und Volksbildung), or Ministry of Education for short. Rust was well positioned in the Nazi hierarchy and had access to the highest levels of government. Spemann opened his letter by reminding him gratefully of the note the minister had sent him on the occasion of his Nobel Prize receipt. He went on to plead von Frisch's case. If he were to be retired, Spemann warned, "German biology is threatened with one of the most devastating blows that could be dealt it." He emphasized von Frisch's international reputation, his tireless efforts to educate physicians, teachers, and scientists, and his centrality to the continued flourishing of the Zoological Institute in Munich. "He is one of our best," he urged, "and without question irreplaceable."[5]

Spemann's letter was emblematic of one kind of appeal on behalf of von Frisch: a famous scientist arguing for the excellence of von Frisch's work and its importance to German science. Von Frisch and his allies managed to mobilize some of the most respected and influential members of the scientific community; these were, indeed, "very decisive people," as von Frisch had hoped in his first letter to Koehler after finding out about the regime's intentions. And yet, as we shall see, the kinds of people who turned out ultimately to be most successful in arguing on his behalf would be drawn from a different, far more applied group of practitioners. Over a year after von Frisch received the initial news, his friend Koehler wrote with some exasperation that the Ministry of Education seemed indifferent to arguments made on behalf of pure science and was willing to "yield only to the pressure of the applied sciences."[6]

Although historians of science have rejected the notion that only the applied, most crudely ideological, or war-oriented sciences were able to thrive in the Nazi state, von Frisch's case suggests that, beyond the most secure and established scientists, making a strong argument for the practical implications of one's work promised the most support, especially as the war dragged on. When von Frisch started to understand that the pragmatic was going to outweigh the theoretical, his strategy shifted decisively on two levels. First, he would recruit more allies from practical fields. Second, he refocused his work from largely academic to practical concerns. This change was not only at the level of public relations but indeed encompassed his research agenda as well.

On February 2, 1941, only weeks after von Frisch had learned that his position was in jeopardy, he published an article in the Nazi publication *Das Reich* that stressed the useful and applied nature of the work pursued in his laboratory. The weekly newsmagazine covered politics, sports, and culture and served as a mouthpiece for Nazi ideologues with regular front-page editorials by the minister of propaganda, Joseph Goebbels. Titled "German Research in the War," von Frisch's piece presented readers with a carefully crafted message about the relevance of his institute's work to practical concerns about food, agriculture, and the war.[7] The article highlighted the wide range of work being undertaken at the lab on issues that ranged from auditory perception in bats to the sense of taste in snails. At its heart, the essay was an effort to introduce the lab's work to a broad German readership and thereby raise its public profile. At the same time, it emphasized the usefulness of the work to the German war effort. Von Frisch knew that the regime was skeptical of pure science for the sake of science. If he were to have any chance of getting the decision to retire him overturned, he would have to make a strong case for the importance of his work to an audience that reached beyond a narrow group of peers.

References to Germany and the war were everywhere in the text and images. One doctoral student, pictured bending over a microscope, wore his military uniform under his open lab coat. The text explained that he had been furloughed from military duty to complete his work on the auditory sense of fish. Another caption informed readers that the young woman pictured observing slugs and snails had been released from her duties with the air alert service to conduct research that would protect German agriculture from the creatures' destructive assaults.

Von Frisch himself was pictured standing by a fish tank while a woman seated next to him took notes. He was studying what happened when a school of minnows, "those little defenseless peaceful fish," were exposed to water infused with the scent of their "enemies," the pikes. According to the caption, when such water was added, the entire swarm reacted immediately and simultaneously by fleeing to the other side of the tank. The alleged goal of the work was to determine what controlled the startling simultaneity of this panic reaction in the animals. Von Frisch explained that the work had practical applications, this time in human psychology, remarking: "It is not a large step from sensory physiology to

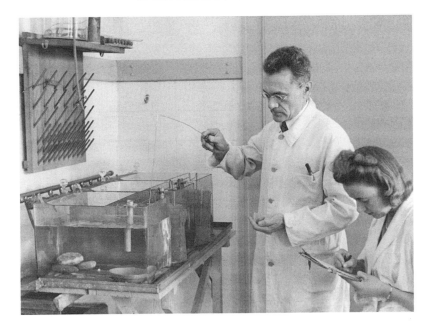

FIGURE 4.1. A photograph of von Frisch and an assistant working on the scare reaction of minnows when exposed to the scent of their "enemies." The image was included in von Frisch's 1941 article for the Nazi publication *Das Reich* that showcased the research being done in his laboratory. (Ullstein Bild/Getty Images.)

animal psychology. . . . Many drives and emotions of the human psyche already appear in very rudimentary form in the mental life of animals." His readers undoubtedly felt the relevance of this work as they were steeped in the rhetorical duality of combat valor and cowardice. Examples of German heroism and enemy cravenness were common propaganda tropes in the early 1940s. But as so much of von Frisch's prior work showed, even when writing for a largely psychological audience, he generally stopped short of making such leaps from animal to human and often cautioned against what he considered to be facile extrapolations between them.[8] Yet in this case, he encouraged the connection.

The article in *Das Reich* also described the von Frisch lab's research on the effects of withholding food from flour moths. Although von Frisch was vague on the results, undoubtedly because the work itself was underdeveloped, he stated: "When the food is offered to them in slightly insufficient quantities, certain effects manifest themselves over the course of generations." He could assume this work would resonate with a readership focused on food shortages during wartime.[9] He also noted that "one can't conduct such experiments with humans." In retrospect and in light of the Nazi atrocities—indeed, large-scale starvation experiments on camp inmates were taking place at precisely this time—von Frisch's statement seems disturbingly oblivious.[10]

With this article, von Frisch made a concerted effort to counter notions among party officials and the public that his was a lab of esoteric animal gazers. While some of his research might not seem obviously useful, he argued, the continued health of science depended on it. In addition, many projects initially undertaken strictly for academic purposes could eventually yield unexpected practical benefits. For example, he informed readers that while the science of heredity had been taken up for purely scientific motives, today the "German Reich seeks to exploit [it] to the widest extent possible for the benefit of the Volk."

Von Frisch's claim that genetics had been taken up for purely academic reasons is of course false, as its history is deeply entwined with concerns about plant and animal breeding.[11] But by invoking eugenics, von Frisch could be reasonably sure that his words would resonate with the readership of the ultra-Nazi publication.[12]

The subject of eugenics was not new to von Frisch or to most scientists at the time. In a popular biology textbook published in 1936, several years before his own heredity was put in question, von Frisch had dedicated an entire section to how genetics from plant and animal breeding might be profitably applied to better the human race. The book had appeared in two editions—the first with the publishing house Ullstein and the second with the Deutscher Verlag under the auspices of the Goebbels Fund for the German *Wehrmacht*, which published reading materials specifically intended to entertain and edify the troops.[13]

As was common for such expositions, in his textbook chapter von Frisch worried that the human stock was declining. To his mind, medical science and human sentiment had largely exempted humans from the merciless selection that acted on the rest of the living world. In addition, the lack of reproductive restraint by those with mental and physical defects and the countervailing desires of the "valuable segments of the Volk" to limit their offspring coupled with the dysgenic effects of war all contributed to the overall decline of the race.[14] He lamented that education alone was not enough to overcome these troubling tendencies and offered as an effective solution the "pain-free operation" that could sterilize those who were "mentally deficient or suffer[ed] from certain heritable mental illnesses or physical malformations." "Naturally, this is a forceful intrusion into the self-determination of man," he mused. He also acknowledged that such measures would likely also result in a collateral deterrence of potentially healthy births, as "defective" parents often gave birth to perfectly sound children. Nonetheless, he deemed the common good to outweigh the potential costs to individuals: "Is it not better that a few fit ones remain unborn," he asked rhetorically, "than that those unfit for life are placed into the world with abandon as torments to themselves and others?"[15]

In the early twentieth century, eugenic ideas were widespread among scientists, politicians, and members of the general public of varied political leanings. Although Germany would emerge as the ultimate example of eugenic horror, aspects of the science were practiced in many countries in the first decades of the century, including the United States, Britain, and Scandinavia.[16] Eugenic policies included such wide-ranging directives as attempts to encourage the healthy or wealthy to have more children, reduce infant and maternal mortality, or demand involuntary sterilization of those deemed unfit.

But the Deutscher Verlag edition of von Frisch's biology textbook contained an additional section that explicitly aligned the discussion with the policies of the Nazi state. Here von Frisch identified four European races: the Nordic, Mediterranean, Dinaric, and Alpine. While he presented these as fundamentally distinct from one another, he qualified that they did not actually exist in pure form, as there had been much intermarriage and procreation. In animals, he argued, such crosses between "races" often gave breeders lots of variations that could, in turn, be used for further crosses to increase the animals' overall quality. At the same time, when crosses yielded progeny with unfavorable traits, breeders could "weed them out" to ensure the overall health of their stocks. In humans, however, there was no breeder who guarded against the propagation of unfavorable results: "nobody is given the right and power to cull the unfortunate types from the products of human race mixing." Of course it was precisely this kind of "right and power" that the German state had and would increasingly arrogate and wield. Troublingly, in this edition von Frisch also added that in place of a breeder, the German Race Laws, the National Socialist Office of Racial Policy (Rassenpolitisches Amt), and the National Socialist People's Welfare Organization (NS-Volkswohlfahrt) had all been charged with overseeing the purity and well-being of the German people.[17]

Von Frisch made no explicit mention of the Jewish race in his discussion of eugenics and declined to offer any relative valuation or hierarchy of the four European races he listed. Nonetheless, the discussion was sufficiently favorable toward eugenic policy to have been deemed appropriate for the Goebbels collection. The publication in *Das Reich* was less explicit about eugenics but nonetheless lent itself to a pro-German reading with its explicit references to serving the needs of the Reich at war.

When Koehler read von Frisch's essay in *Das Reich*, he wrote to him to express his delight with the text and photographs. "It is all so lovely," he enthused. "That this is published in *Das Reich* really just does not square with what one hears about how you are being treated."[18] Koehler's comment invites speculation about how von Frisch managed to place this deliberate staging of his work and institute in such a high-profile Nazi publication after receiving news of his non-Aryan descent. While the details surrounding the publication remain unknown, its venue underscores that von Frisch and his allies targeted and often succeeded in accessing high levels of the Nazi bureaucracy.

At the same time that efforts to rally supporters were underway, he also frantically sought to find out more about what exactly the authorities in Berlin had unearthed about his background. The notice of his imminent dismissal had only stated that he had been found to be one-quarter Jewish. The specific details surrounding this determination were not included, and nobody knew initially whether this meant that one or two of his maternal great-grandparents had been Jewish. In the perverse accounting of the Nazi Party, such details mattered critically and would help von Frisch determine what the best course of action might be. Koehler summed up the situation: "If both great-grandparents are non-Aryans, then one ought to plead, [but] if it's only one, then one must demand."[19]

In addition to the uncertainty of his descent, nobody knew if or when the call for his retirement would be submitted. To complicate matters, if such a request were to be made, it would be of some benefit to von Frisch to exploit any bureaucratic lag and preemptively submit his own request for emeritus status. Whether the retirement was voluntary or forced by the state would most likely have repercussions as to whether he would be allowed to continue working, at least on a reduced scale, or whether he would be forbidden from pursuing his research altogether. The key, then, was to submit his own request as soon as it was clear that the official proceedings would go ahead, but no sooner, so as not to foreclose the possibilities that the call for retirement might simply linger unfiled for an indefinite period of time.

While von Frisch waited for the Ministry of Science, Education, and Public Instruction to send him information about the nature of the incriminating documents in Berlin, he contacted his oldest brother, Hans. His three brothers in Austria, as well as a cousin at the University of Munich, had been allowed to continue in their civil service positions more or less undisturbed. But Hans's example might offer further insight; he had joined the Nazi Party in Austria under the Catholic regime of Engelbert Dollfuss. Because involvement with the party had been illegal in Austria prior to the country's annexation by Germany in 1938, he had briefly lost his job at the Technical University in Vienna. After 1938, he was reinstated but left the party two years later. Although he "didn't like to talk about it," as Karl von Frisch remembered, his leaving the party in

1940 had not been entirely voluntary and, as such, might provide further details about the ancestry of their maternal grandmother.[20] Now von Frisch wondered what Hans might know of their lineage.

Hans replied to his brother's query in a letter that explained that he, too, did not know the exact nature of the evidence that had been compiled against him, but he recalled the events that led up to his leaving the party.[21] Shortly after the *Anschluss*, a municipal office had requested him to appear before them to present his papers. When Hans arrived at the office, an official quickly paged through his documents to make sure they were complete. As he leafed through the pile, he remarked on the last name Exner. He also informed Hans that the reason the papers had been requested was that officials were considering him for a position on the Faculty of Law at the University of Vienna.

After he delivered the papers, Hans heard nothing further about the position or his candidacy. The following fall, the Reich Genealogical Office also requested to see his documents. Shortly after, Hans was ordered to renounce his Nazi Party membership because of his compromised ancestry. Then in June of 1940, the Ministry of Education let him know that no further action would be taken against him if he submitted his own request to be relieved of his university duties. But the president of the university discouraged Hans from submitting the request and encouraged him to wait. Hans, however, decided to go ahead and file for his release. A few months later, he was officially suspended from his post. But at the same time that the ministry passed down the decision, the president once again took Hans's side and asked him to continue with his teaching duties. Thus, in the end little changed for Hans; he maintained his salary and continued working despite having been officially relieved of his duties.

The case underscored the considerable leeway with which the designation of Quarter Jew might be handled. But it also confirmed for von Frisch that his brother had been asked to leave the party because of his Jewish descent. This realization did not bode well for his own situation. When Koehler learned that the regime considered von Frisch's brother Hans one-quarter rather than one-eighth Jewish, he despaired: "That I fall asleep and wake thinking of you doesn't help you at all. In opposition to the hardness of these laws, no conventional perspectives or ways of arguing can shatter anything."[22]

But after an initial flurry of calling in favors and pulling any strings that seemed safe and worthy of pulling in Berlin, von Wettstein and Kühn got lucky—on February 5, 1941, they managed to get an appointment with the undersecretary of the Ministry of Education, Franz Kummer. Once in the office, the men explained von Frisch's situation. They emphasized that to lose him from their ranks would deal a blow to German science that it could scarcely afford.[23] The minister seemed interested and impressed by von Frisch's accomplishments but finally told them that he regretted that the case was not his to decide. He instead referred them to Max Demmel, another senior official in the Ministry of Education.

Demmel, too, appeared friendly and attentive as the two men laid out their plea. Yes, he assured them, he had heard of von Frisch's case and was aware of his significance to German science. He had even read the Nobel laureate Spemann's letter to Minister Rust. But he too regretted that the matter lay outside his jurisdiction. He had "processed the case according to duty" and passed it along to the office in charge of legal matters pertaining to the civil service. He was afraid that there was really no way to stave off the inevitable ouster save by an act of mercy by the Führer.

"My Dear Karl," Kühn wrote later that evening to von Frisch, "I am so sad. . . . I would so much have liked to be shown a way out that I had not yet seen. And I still hope that something might be done. But the hope is not very big. It is now in the hands of the bureaucrats and in the party machinery and I see no way by which we might pull it out from there."[21]

Then on February 13, 1941, von Frisch finally received a copy of the document from the Ministry of Education in Berlin that proved that his maternal grandmother had been Jewish, according to church sources. She had been born and baptized on October 3, 1814. But her father and mother, it turned out, had been baptized just over a decade before their child's birth. The Catholic Church in Vienna had supplied the relevant records.

With this, von Frisch's ancestry was clear—like so many Jews of the early nineteenth century, his great-grandparents had converted to Christianity in the hope of leading an assimilated life as citizens of the Austro-Hungarian Empire. But this didn't matter to the Nazis, whose concept of what it meant to be Jewish was overwhelmingly based on blood and descent rather than cultural heritage or religious belief. As a consequence, von Frisch, who had never before considered himself Jewish, was now officially declared such more than one hundred years after his great-grandparents had converted. The cover letter that accompanied the document demanded that he submit an official statement of his position vis-à-vis his ouster within eight days.[25]

Now with his non-Aryan status confirmed and the clock ticking, von Frisch and his allies in Berlin (especially the biologists Kühn and von Wettstein) moved into high gear. They only had eight days to determine whether an exception might be made on von Frisch's behalf. In the meantime, they also learned that he would not qualify for voluntary retirement because he was under the required age cutoff of sixty-two. The only alternative to accepting or submitting a request for retirement was to appeal to the Ministry of the Interior on the grounds that von Frisch should be "equalized" (*gleichgestellt*) to someone who had either fought on the front in World War I or been called into the civil service prior to its outbreak. Recall that the 1937 Civil Service Law had made exceptions for non-Aryans on those two grounds. However, such an appeal would have to pass through the highest ranks of the party, and it was very doubtful that the request would be granted.[26]

The situation was grim and the deadline fast approaching. In consultation with his allies, von Frisch decided to travel to Berlin on February 18 for two days. Once there, he spent the days in meetings with colleagues and officials who might be in positions to help.[27] One of his first visits was to the well-known surgeon Ferdinand Sauerbruch, one of the most powerful men in German science. Like the universities themselves, science under the Nazis was reorganized according to the *Führerprinzip*. As a consequence, individual "leaders" had almost complete control over funding and personnel decisions within their divisions.[28] In 1937, the Ministry of Education had founded the Reich Advisory for Research (Reichsforschungsrat, or RFR) in an effort to centralize science planning and funding. The RFR was divided into subject areas, each of which had a scientist at its head. Starting in 1937, Sauerbruch had been placed at the head of the Medical Division.[29] Although he never joined the party, he had critical access to Werner Zschintzsch, the state secretary of the Ministry of Education.

Von Frisch also met with Wilhelm Führer in the Ministry of Education.[30] Recall that Führer had in 1936, as the leader of the Instructors League, written a scathing letter in which he demanded that inquiries be made into von Frisch's ancestry. Von Frisch, of course, was unaware that the very person to whom he now made his appeal had written such a letter. Führer assured him that nobody in particular was creating trouble for him. Rather, the request to retire him had been automatically generated through his brother Hans's dismissal from the party. Führer regretted that exceptions were granted only rarely in cases where people had rendered the party extraordinary services. Unfortunately, the conditions at von Frisch's institute would hardly qualify him for such an exemption; Führer reminded him that in the early days of Nazi rule, he had employed a suspiciously high number of Jews. On the other hand, he assured him, his considerable scientific achievements were known, and the ministry did not wish to disturb this important work.

Later in the day, von Frisch also met with another senior official of the Ministry of Education, Max Demmel, with whom Kühn and von Wettstein had previously spoken about him. While Demmel did not offer him any concrete cause for hope, he did extend the deadline by which von Frisch would have to reply to March 10. This was a welcome reprieve and would give him some additional time to weigh options and mobilize support. Demmel also assured him that he was aware of the importance of von Frisch's current research and that the authorities did not wish to hinder the progress of this work. However, he criticized him for not having reported his non-Aryan ancestry of his own accord. Von Frisch protested that he had, in fact, appended a statement to his questionnaire in 1937 in which he alerted the authorities of his grandmother's uncertain provenance. He worried that somehow this information had not made its way to Berlin. Demmel now advised that if he could prove that he had made such a declaration, it would certainly reflect favorably on his case. Von Frisch assured him that he would try to locate the statement and have its content communicated to Berlin immediately.

All in all, the visit left von Frisch hopeful. He was especially pleased that both Führer and Demmel conveyed that the National Socialists recognized the importance of his work and did not wish to interfere with it.[31] He also left with the hope that Sauerbruch would be able somehow to intervene on his behalf with State Secretary Zschintzsch. He wrote a coded message to his longtime publisher Ferdinand Springer of this possibility by obliquely referencing Sauerbruch's vocation: "I received an extension without problems and by the time the deadline is up, there might even be an operation with smooth healing on the part of the surgical contingent."[32]

By the time the new deadline approached, von Frisch was buoyed by assurances from various sides that his retirement would most likely not be pursued, and if it were, he would be given enough notice to preempt and stall the request with an "equalization" application of his own. He finally composed a confident reply to the president of the university on March 11, 1941.[33]

In the letter, he acknowledged that he had been counseled to submit a request for his retirement of his own accord. But he rejected this option on the grounds that he was "not aware of having ever in word or deed done anything that would have run counter to the interests of the National Socialist State." "On the contrary," he continued, "I believe, that I have and may continue to serve the state to the best of my abilities and strengths through my teaching, scientific work, and not least my efforts to disseminate scientific knowledge to popular audiences." He closed the letter with a "Heil Hitler!" above his signature.

By June of that year, no negative news had reached him, and he felt optimistic that his own efforts and the behind-the-scenes maneuverings of sympathetic colleagues had done the trick. He wrote with relief to Spemann to let him know that all seemed well. He had received explicit assurances from his "main advisor" (presumably Sauerbruch) that Zschintzsch had assured him the retirement would not go forward. Spemann, in turn, expressed his happiness at the news and concurred: "After all this, I don't think it will be possible for enemy powers to cause any more trouble. Now you can continue your beautiful work with a light heart. . . . You are probably already dreaming of being in Brunnwinkl."[34] Spemann did not live to see the outcome of von Frisch's situation, as he died just a few months later in September of 1941. But for the time being, there was cause for optimism.

Koehler was also pleased when he wrote to von Frisch. But he also expressed surprise. Based on what von Frisch had written him, it seemed that important bee work had somehow saved the day.[35] That he would secure a reprieve because he was a successful teacher or because the institute would not continue properly without him, sure, but why would anybody in the Ministry of Education care whether he continued his work on bees?

Koehler's question about the twist in von Frisch's fate was justified. That the Nazi government would even have taken such an interest in von Frisch's work attests to the desperate state of German apiculture in 1941. Indeed, all was not well with the insects.

In 1940 and 1941, beekeepers began to observe alarming declines in their populations due to the intestinal parasite *Nosema*. *Nosema* is generally found in bees, but at certain times the microorganism population spikes, causing massive numbers of deaths among the insects. Afflicted bees are wracked by diarrhea, weakened, and incapable of flight before they die. By von Frisch's own estimates, the plague destroyed 800,000 colonies (with each colony containing 30,000–50,000 worker bees) in 1941, its worst year.[36] Then as now, such a calamity made visible what otherwise went largely unremarked in postindustrialized societies—bees ensure the propagation of most of our fruits and vegetables. And in times of crisis, their role comes painfully to the fore: a large-scale loss of the pollinators spells catastrophe for animals and humans. This prospect was especially frightening to a nation engaged in a war of attrition and teetering on the brink of food insecurity.[37] Because of this imminent danger, the Ministry of Food and Agriculture founded the Nosema Task Force.

In the summer of 1941, just a few months after von Frisch received the news of his imminent retirement, he began to investigate the problem of *Nosema*.[38] This was a drastic shift from his earlier work. Where previously his research had been guided largely by intellectual curiosity and focused on practical and applied problems only secondarily, he now dedicated his entire lab to finding a solution to the question of the bee deaths. This meant that a majority of his bee stocks, as well as students and assistants, were put to the task of determining a solution to the ominous plague.

In his autobiography, von Frisch recounted, "The head of the [Nosema] task force was a man who was influential in the area of food. He knew my work and that I was threatened in my position. He argued emphatically on my behalf and that of our institute. It was because of him that I received the work commission from the Ministry of Food and Agriculture to investigate the *Nosema* plague threatening the bees."[39] Von Frisch was tight-lipped about the identity of the man in charge of the task force, presumably to protect in the postwar period the man who had wielded such influence in a Nazi ministry. But fleeting traces in publications, letters, and especially a denazification testimonial by von Frisch suggest that the person in question was a physician by the name of Franz Wirz. Wirz was in charge of the Main Office for Health (Hauptamt für Volksgesundheit). He was responsible for food planning and, in this capacity, oversaw the Nosema Task Force for which he was able to recommend von Frisch.[40]

For von Frisch, the commission provided access to scarce resources and presented an opportunity to officially reorient his work. As a result of the contract, he received funding as well as additional personnel to help him with the investigations. He also obtained permits to travel as necessary throughout the Reich and was issued precious gasoline rations. Significantly, some of von Frisch's lab members were given furlough from active combat to help pursue these investigations.[41]

During the summer of 1941, von Frisch and his collaborators began to test in Brunnwinkl the efficacy of conventional methods of fighting the disease as well as new chemical methods. In addition, they established a network of observers throughout Germany that were to report on any local environmental conditions that might hinder or promote the disease. Although the results of the work were largely inconclusive in terms of solving the problem of *Nosema*, they would have profound consequences for von Frisch's fate.[42]

Shortly after beginning his investigation of the disease, von Frisch also obtained permission to expand the focus of his research to the problem of crop pollination. Since bee populations were under siege due to the intestinal parasite, it became all the more imperative that the diminished stocks pollinate sufficient flowers to ensure crop yields and food supplies. Recent reports from Russia had proclaimed that bees could be scent trained to increase their visits to nearby crops. The Reich Specialty Group Beekeepers had a book on this topic translated so scientists could test the Russian claims and make recommendations as to the feasibility of this technique for large-scale adoption throughout the Reich. Later von Frisch explained: "It is to be expected that one can achieve several goals through purposefully designed scent feeding: the quick approach of [a bee to] a particular plant, a quantitative increase of approaches, an increase in the bees' work intensity, and a lengthening of their work time."[43] It was of course in von Frisch's interest to emphasize the likely fruitfulness of his investigations as well as the need for future trials. After all, his livelihood now depended on it. That he expressed these in terms of raised "work intensity" and "lengthening of their work time" indicates that he viewed the bees largely in terms of their labor capacity in these studies. This was the moment when von Frisch cemented his commitment to harness the bees into the service of the Reich to contribute their part to the war effort.

But by late summer 1941, events had once again taken a turn for the worse. On August 25, Kühn wrote to von Frisch in Brunnwinkl, "I have just heard some very alarming news from a source I cannot reveal: Your situation either did not take after all or has once again been stirred up somehow. It appears urgently necessary that you put in an 'equalizing application' with the Ministry of the Interior so that the situation is not suddenly decided negatively on purely bureaucratic grounds." Kühn regretted that he had no further information and therefore felt hesitant to act. He also wondered whether someone in Munich had once again agitated against von Frisch.[44]

The worry that someone in Munich might be causing these troubles nagged at von Frisch and his supporters throughout this period. He wondered early on whether there might not be someone acting on a personal vendetta against him. Assurances to the contrary from party officials in Berlin notwithstanding, these suspicions were indeed warranted.

The president of the National Socialist Lecturers' League, Ernst Bergdolt, had gotten wind the previous spring of the possibility that von Frisch's retirement might not proceed. He dashed off a vitriolic letter to Demmel of the Ministry of Education in Berlin.[45] Demmel had of course met previously with von Holst, Kühn, and von Frisch and was by now familiar with the situation. Bergdolt's letter expressed perhaps the nastiest and most petty denunciations that survive in von Frisch's file; it even included a list of "suitable" candidates with whom to replace von Frisch after his ouster. According to Bergdolt, supporters unjustly praised the scientist as being "the best, unique, and irreplaceable" among German zoologists. "Interested opposing parties have succeeded with all means of propaganda and personal influence to create a psychosis, to which even some National Socialist persons in good faith have fallen victim." Because of his suspect "political behavior," von Frisch was "in no way deserving" of an exemption from the law. Indeed, those who praised his international reputation ought to bear in mind that his institute had been funded through "American money," and that he "did not disappoint his money givers, since whenever in any way possible, he favored and encouraged Jewish scientists and crossbreeds."[46] The letter went on to enumerate the Jewish colleagues whom von Frisch had tried to help as the situation throughout Germany became increasingly desperate. It faulted him for not only encouraging non-Aryans but also for actively discriminating against National Socialists. The "excellent Strasburg zoologist [Ludwig Döderlein], whose scientific importance von Frisch could not even begin to approach," was mistreated by being "shoved off into a small, poorly heated, and completely inadequately furnished room, while the Jews distributed themselves among the nicest and best-equipped rooms of the new institute." According to the letter, Döderlein suffered von Frisch's maltreatment for no other reason than that he was not part of the lab's preferred "philo-Semitic or, better yet, Jewish-blooded clique." If von Frisch were to be permitted to continue working at the institute—and this was a big *if*, as the author could not endorse such leniency—he should himself be confined to the pitiful space that he had assigned the Nazi emeritus Döderlein. Bergdolt's phrasing of the letter and his points of emphasis leave little doubt that he had read or at least heard of Spemann's letter to Rust on von Frisch's behalf. In any case, Bergdolt's denunciation had gathered sufficient steam by late August that Kühn was put on renewed alert about von Frisch's situation.

After receiving Kühn's letter, von Frisch at once called his daughter Maria who had stayed in Munich to work for the summer and was therefore closer to the machinations against her father. He now asked her whether she had learned anything concrete about the rumors of his renewed troubles. She informed him that most of his confidants in Berlin were out of town, and that nobody really knew whether anything new had been generated or whether these were simply recirculated, old rumors.

Since the Ministry of Education would make the final decision, it was imperative now to get in touch with Zschintzsch. Von Frisch fretted to Kühn that everyone was on vacation and that he did not know how best to put in the equalization application. Since even a clerical misstep could have dire consequences, he asked Kühn to get in touch with Zschintzsch's cousin Dr. Crampe, whom one of their acquaintances knew, in the hope that he could connect them with Zschintzsch. There was another man who might help, but in his agitated state, von Frisch could not remember his name, only that he had spoken to him with von Wettstein during his time in Berlin. He asked Kühn to please alert him by telegram as soon as he learned more with either of two messages: "Submit application" or "Do not submit application."[47]

The Munich physicist Arnold Sommerfeld had also expressed concern, this time to Sauerbruch, to inform him that von Frisch's case seemed once again to have been "stirred up," and that the situation appeared in a "dangerous state." Von Frisch, he added, had been advised to put in an application for "equalization," a move he himself deemed "rather desperate." Sommerfeld asked Sauerbruch whether he would "once again do [him] the great kindness of trying to find out what lies behind the rumor by going directly to State Secretary Zschintzsch." He asked that he try to determine "whether there was something to the rumors, whether he [von Frisch] is in imminent danger, and whether he should put in the request for equalization." Four days later, Sauerbruch wrote a cryptic one-line note to Sommerfeld: "Dear Colleague! In all haste a message that I have acted in accordance with your wish."[48]

At around this time, von Frisch's old friend Otto Koehler urged him to scour his list of former students to see if he could find anyone from outside academia who might be able to help. "Sometimes," he wrote, "a random Frau Beelady is very important." Koehler felt strongly that such people with ties to practical beekeeping might be helpful in von Frisch's quest to retain his position, as he deemed them "better advocates than dry professors."[49]

Just such a supporter was Herr Schreiber, the president of the Agricultural Council for Beekeeping of Upper Bavaria. Schreiber sent two letters on behalf of von Frisch, one to the surgeon Sauerbruch and the other to Gerhard Klopfer, the state secretary in the Nazi headquarters in Munich known as the Brown House. In the letter to Sauerbruch, he urged that "all German Beekeepers would be extremely grateful if you were to advocate for the preservation of the most successful bee researcher of the world." To Klopfer he wrote: "As a representative of practical German apiculture, I allow myself, Mr. Secretary, to ask you to do all that it is in your considerable personal power to save us this irreplaceable researcher." The letter continued with an impassioned plea on behalf of the dying bees: "It ought to be known to you, that the Führer, as a result of the catastrophic emergency situation befalling German beekeeping and as a result of his understanding of the importance of apiculture, which he inherited from his father, has in the last two years given decisive orders to support and encourage German beekeeping. Therefore, the removal of this important bee researcher runs counter to the intentions of the Führer."[50] The reference to Hitler's beekeeping sympathies through his father pointed to the fact that the latter was known to have kept bees.[51]

Thus, in addition to their early strategy of rallying the brightest and best-known scientists, von Frisch and his allies began to pursue this parallel tack of appealing to a less academically distinguished but ultimately more successful set of supporters for help. The strategy would prove astute, as the power base in German science and politics had shifted under National Socialism. Basic research was never abandoned under the Nazi regime, but greater emphasis was placed on applied sciences that either served Hitler's political and racial ideologies directly or contributed to the war effort. This latter focus was not unique to this nation at war.[52] However, in the case of Nazi Germany, the transition to a more practical and agriculturally focused science was also in keeping with the regime's explicit ideology of blood and soil (*Blut und Boden*). From the search for cutting-edge weapons to the mass-scale killings in the camps, Germany embraced the products of modern industrialization for its murderous ends. At the same time, a strong cultural current praised a simple, closer-to-the-earth German existence.[53] This ideological movement gathered its steam through the creation of founding myths that referenced an unseemly mix of historical kitsch and nostalgia. In these narratives of proto-German valor, the Aryan race was celebrated and recast as being at one with the land.

This ideology meant a boon to those areas of scientific research that focused on the land. Entomology was one of the five special scientific divisions Heinrich Himmler led as the president of the Ahnenerbe. The Ahnenerbe was the scientific branch of the SS originally founded in 1935 for the purposes of conducting research and providing instructional material on historical and *völkisch* topics that promoted the ideology of the regime. Under the auspices of the SS, research was generally racist and often deadly. Researchers working for the Ahnenerbe also maintained a strong interest in plants, foreign lands, and insects. While investigations in these areas might at first glance appear harmless, support of these fields of research was undergirded by fantasies of German expansion and autarky.[54]

In 1942, Himmler founded a new entomological institute that was aimed at studying and eradicating insect pests. Ironically, von Frisch had been briefly floated as a contender for its leadership but was soon dropped on the grounds of his Jewish ancestry. The institute fit well with the German view of nature that cleaved the world into a binary division of *Schädlinge* and *Nützlinge*. The terms literally translate as "harmers" (or "pests") and "helpers." The goal of the institute was to study the life habits of flies, lice, fleas, and mosquitoes in order to come up with ways of eradicating them. In addition to devising new chemical means, scientists were to determine "through which plagues and bacteria we humans can bring about and encourage the destruction of these damaging insects." In addition, Himmler ordered the investigation of how larger animals (such as birds, mammals, and snakes) might also be enlisted to destroy the broods and larvae of those insects.[55] Himmler declared nothing short of chemical and biological war on the hated creatures.

But not all insects were deemed bad, as the dichotomy of the "harmers" and "helpers" suggests. If the "harmers" were entered into the ledgers of agriculture and forestry in red ink, the honeybee was decidedly captured in black.[56] With its hard work and communal sacrifice to the common good of the Volk, the honeybee appealed as both an insect valuable to agriculture and a useful cultural referent for human society. And as previously mentioned, the crisis of the bees was deemed serious enough for a special division to be formed in the Ministry of Food and Agriculture to investigate the problem.

Whether Otto Koehler consciously tapped into this pro-rural cultural and political resource is unclear. But his instincts were sound when he urged von Frisch to scour his list of former students to see if he could find anyone of influence other than "dry professors."

At about the same time that the beekeeper Schreiber wrote his letters to Sauerbruch and Klopfer, Koehler managed to mobilize another man who would soon make headway on behalf of von Frisch, Bernhard Grzimek. Grzimek would become the keeper of the Frankfurt Zoo and a well-known television personality in postwar Germany.[57] He was a veterinarian by training and had been a functionary in the Ministry of Food and Agriculture since 1938. His wartime work focused on the control and eradication of cattle and poultry plagues. While Grzimek was not of the same scientific stature as many of von Frisch's other supporters, he had critical access to the Ministry of Food and Agriculture and the party. As a consequence of Grzimek's maneuverings, Koehler marveled to von Frisch that the Ministry of Food and Agriculture was the only body to have made inroads with the leadership in charge of von Frisch's case.[58] Here it was again—those with connections to applied fields that related more or less directly to the Reich's agricultural and, therefore, food needs were in a strong position in these times of total war.

And finally, on February 26, 1942, there appeared a sliver of hope for von Frisch's position. He received a letter from the Bavarian Ministry of Education and Culture in which he was granted permission to "continue his research to combat the *Nosema* plague in bees in his current institute after his retirement."[59] This still meant that the ministry would go ahead with his retirement. But it was, nonetheless, a major coup, as it would allow him to continue working on the project most dear to the regime and would give von Frisch continued access to funding.

Still, von Frisch was not satisfied and wasted no time in further urging along his case. The same day he was granted his reprieve, he wrote back to the Bavarian Ministry of Education and Culture. He explained that he was now fully committed to solving the pressing problem of *Nosema*. But, in addition, he wondered whether he might be permitted to expand this research from this narrow focus after his retirement. He explained that his "freedom of research" would be impaired if he was forced to dedicate all of his remaining research life to the plague. He also hinted that since he now was using all resources to address the problem, much of the equipment would be tied up and therefore unavailable to whomever was to take over the institute's directorship after his retirement. Finally, he pleaded once more to stay in his position. If he were to be retired, the students he had selected to work on the project would no longer stand under his guidance but would instead pass to the supervision of his successor. "This working community" he lamented, "is being quite literally uprooted through my retirement, and my research is being disturbed most severely."[60]

This last argument may have come across as disingenuous or at the very least out of touch with the realities of what he was facing. But there can be no question that the uncertainty about what would happen to the community of students and collaborators with whom he had surrounded himself over the years weighed heavily on his mind. Just as the bees depended on their hive mates for survival, so too did his science rely on the close cooperation of the members in his scientific group. In later years, he would remember that it had been critical for him to work with people whose judgment and reliability he trusted. The task at hand was no one-man job and instead depended on the eyes and hands of many.

But despite the validity of von Frisch's claims with respect to his scientific approach, the letter to the ministry was nothing short of remarkable. To ask for further accommodations, and moreover to suggest that none of the lab's equipment would be ceded to his successor, after having just been granted permission to work after his forced retirement must have smacked of either stunning gumption or galling stupidity. But regardless of how the letter was received, for the moment there was nothing von Frisch could do but carry on with his research and await the ministry's response.

In the meantime, Grzimek continued to advocate on his behalf. He again spoke with colleagues in the Ministry of Food and Agriculture and argued for the importance of von Frisch's work to

German agriculture. As a result of these efforts, an official of the
ministry by the name of Georg Narten finally agreed to personally
speak to the state secretary Werner Zschintzsch. This intervention
promised a major breakthrough for von Frisch.

On May 9, 1942, Narten entered Zschintzsch's office to plead von
Frisch's case in the name of the Ministry of Food and Agriculture.
Narten explained to Zschintzsch that von Frisch was one of only
three biologists working in Germany whose work offered any relevance to practical matters of agriculture.[61] His work on the *Nosema*
plague was especially critical. Narten emphasized that von Frisch
had already achieved great successes in this line of work and would
not be able to continue on the same scale should he be retired. Although it was certainly true that the work would be interrupted by
his removal from his post, the results of the research up until this
point had in fact been rather modest. Nevertheless, the issue of the
800,000 dying colonies clearly made an impression, and the fact
that such significant resources in von Frisch's institute were now
being dedicated to the work was critical. Narten also told Zschintzsch about von Frisch's recent work on scent-guiding bees to food
crops. He pointedly added that the Soviets had used this method
to raise their agricultural productivity. Finally, he argued that von
Frisch's position was similar to that of the surgeon Sauerbruch in
that he had countless students who were working in agriculture
and pest control. His work, therefore, was critical to the Ministry
of Food and Agriculture and had to be allowed to continue at its
present scale.

The conversation was a success. Zschintzsch was receptive to
Narten's plea, and before the two men parted, he assured his visitor
that no further action would be taken against von Frisch without
the express approval of the Ministry of Food and Agriculture.

On July 24, 1942, the Reich Ministry of Education in Berlin made
the decision official in a brief missive to the Bavarian Ministry of
Education and Culture. The head of the ministry, Rudolf Mentzel,
wrote: "In consultation with the Leader of the Party Chancellery,
I have postponed the pursuit of Professor von Frisch's retirement
until after war's end."[62] Koehler was elated. Von Frisch's emotions

we can only guess. But surely the sense of relief must have been overwhelming. The decision by the Ministry of Education meant that he would be able to continue his work without fear of further action against him until the end of the war.

A few words about von Frisch's situation in light of the larger events that surrounded him are in order. Stressful though it was, von Frisch's ordeal pales in comparison to that of the millions who suffered and lost their lives under the regime. Von Frisch's life was never in danger, and he was able to maintain his property and livelihood. Indeed, Otto Koehler noted in April of 1942 that, while he himself no longer had any doctoral students or staff, von Frisch's laboratory seemed to be flourishing. Thirty students were enrolled in von Frisch's laboratory course and he was supervising fourteen dissertations. A postwar form filled out for the American military government reveals that his personal income also rose by almost 10,000 reichsmark from 1940 to 1941 and declined only slightly in 1942.[63]

What then do we make of von Frisch in this time of duress? It is difficult to glean his emotions or thoughts during the period. While friends and relatives poured out their sympathies in letters, von Frisch remained largely stoic in his replies. He maintained a productive work schedule throughout and his ability to completely revamp the orientation of his work bespoke his tremendous personal and professional resources in the face of extreme stress.

Nonetheless, it is difficult to shake the image of a scientist who escaped the horrors that surrounded him by burying himself in his work. In his postwar autobiography, he wondered whether it would have been possible to stave off the Nazi takeover of the universities if there had been vocal and united opposition at the very beginning. But, he admitted, "many professors welcomed the changes, some out of caution, others from conviction. And soon it was clear that any serious opposition would lead to one's personal destruction."[64]

# DEEP INSIDE THE HIVE

From 1942 to 1944 von Frisch produced a film called *The Honeybee—Development of the Bee and of a Colony* as part of a two-film project for the Reich Institute for Film and Image in Science and Education (Reichsanstalt für Film und Bild in Wissenschaft und Unterricht, or RWU for short).[1] The film offers viewers an intimate portrayal of how the bee Volk had come to stand in for the human Volk.

As the camera reaches deep into the hive, the queen in extreme close-up dips her abdomen repeatedly into cells to deposit her eggs. From these tiny structures, glossy larvae hatch. These wiggle and grow until the worker bees cover their cells to protect the next phase of their development. The camera shows how they mature within their sealed chambers and finally emerge as fully formed worker bees. Following these scenes on the individual worker bees' development, the film offers a parallel narrative of the hive: the queen lays her eggs, the workers care for the new brood, and a new "Volk" is founded.

Unlike the previous films that von Frisch created to accompany his lectures to students and colleagues, this film was conceived and made for the purposes of general distribution to German schools.[2] The black-and-white film boasts obvious cinematic and editing superiorities to his earlier films. For the scenes that show the developing bees, part of the comb was cut away and replaced with a pane of glass through which the footage was shot, allowing the camera to penetrate the hive's innermost recesses and show the developing animals in their cells. The technical challenges were significant and the completion of the film was delayed precisely because of these scenes.[3]

But beneath the film's gorgeous images and mesmerizing details lies another, more unsettling message that does not become apparent until the final scenes of the film. The last intertitle—"The Battle of the Drones"—announces that all is not well in the hive. A new type of bee appears in these final scenes. Previously the viewer had come to understand the function and activities of the queen and her workers, each useful in her own right to the community's well-being. But now we encounter large, sluggish bees among the workers—the drones. The narrative logic of the film suggests that these grotesque, struggling creatures must be expelled and killed for the benefit of the hive. The final scene shows them writhing on the ground with ants crawling all over them.

No prior background about their use to the hive has been established. The movie does not mention the nuptial flight when drones fertilize the queen's eggs in midair to ensure the continued life of the colony. The omission is telling.

In his popular book *The Dancing Bees*, von Frisch's portrayal is no more sympathetic. In the early sections of the book, he discusses drones in general descriptions of the composition of the

FIGURE III.1. The graphic poem on the bees by the popular nineteenth-century satirist Wilhelm Busch. Here the "gluttonous, fat, lazy, and stupid" drones are shown "still lazing in bed" while their sisters are hard at work tending to the hive. (Wilhelm Busch, *Schnurrdiburr, oder: Die Bienen*, 6th ed. [Munich: Braun und Schneider, 1884], p. 4.)

hive and sex determination. But in a later section, also entitled "The Battle of the Drones," he goes into somewhat more detail about the males' role in the hive. And the portrait is not flattering. Quoting the wildly popular nineteenth-century poet and cartoonist Wilhelm Busch, he wrote, "'Gluttonous, fat, lazy, and stupid,' . . . they do not attempt to take any part in the collection of food, an activity for which they are not properly equipped by nature, anyhow. Most of them are too indolent even to help themselves to their own share of the hive's food stores, leaving it to the worker bees to feed them." He continued, "The brain of the drone is smaller than that of both worker and queen—we are not left in any doubt as to the intellectual inferiority of the male in this case." And perhaps the most damning: "The necessity of fertilizing the queen is the sole justification of the drone's existence—each queen requiring but a single drone for this purpose. And yet extravagant nature produces hundreds and hundreds of drones in each colony fated to perish again without having ever gained their object in life; a fate which they share with many a living creature."[4]

Despite von Frisch's acknowledgment that the drones "are not properly equipped by nature" to partake in food collection, he cast this failure in moral terms. The drones are, as Busch suggested, "gluttonous, fat, lazy, and stupid." As such, they offended von Frisch's sensibilities about nature. He pointed out again and again in his writings that nature does not create in vain and that her beauty may best be appreciated through the wonderful adaptations that abound, especially in the animal kingdom.[5] The drones hold an awkward position in this worldview, and von Frisch conceded that rather than wise and parsimonious nature, we are here confronted with "extravagant nature." As if turning on this egregious condition, the hive itself revolts against the abomination of the drones. Von Frisch closed the section with reflections on what he deemed an inevitable massacre: "Thus they [the drones] find their inglorious end at the portals of the bee dwelling, driven out and starved, or stung to death, on a fine summer's day. And this is the meaning of the 'Battle of the Drones': not a sudden upsurge or a 'Massacre of St Bartholomew,' as some poets writing of bees would have it; but a slowly rising hostility on the part of the worker bees, which may drag on for weeks, getting fiercer and fiercer all the time, until every single drone has been killed. From that time onwards until the following spring, the females of the colony, left to themselves, keep an undisturbed peace."[6]

In his written work, then, von Frisch described the worker bees' turning against the drones in rational, even sympathetic, terms. The steadily rising resentment finally gives way to action—the drones are expelled and pushed to their death and the "undisturbed peace" is once again restored.

But the film version of the battle of the drones provides no such context and may be read on yet another register. The film was produced between 1942 and 1944 at the height of the Nazi campaign against the Jews. Historians of Nazi Germany and film have shown motion pictures as favored propaganda tools of the German Reich with production reaching unprecedented highs during the war years under the minister of propaganda, Joseph Goebbels.[7] The early 1940s also saw the production and release of some of the period's most notorious anti-Semitic propaganda films—such as Veit Harlan's *Jud Süß* (Jew Süss) (1940), Wolfgang Liebeneiner's *Ich klage an* (I accuse) (1941), and Fritz Hippler's flat-footed hate film *Der ewige Jude* (The eternal Jew) (1940). Although the films of the RWU never came under Goebbels's direct control (as only sound films fell under his purview), the bee film's righteous expulsion of the swarthy parasitic drones evokes vivid parallels to the anti-Semitic visual rhetoric of the period. Jews were often cast as parasites, which fed on the state body and needed to be hunted down and eradicated to restore the health of the nation.[8]

Thus, in this film of the 1940s, the scientist and his science had disappeared from view in contrast to von Frisch's earlier films. Instead, close-ups broached the confines of the hive to lay bare its most intimate domestic scenes as well as the chilling slaughter of the drones. These two films made in collaboration with the RWU marshaled cutting-edge technology that arose from an increasingly sophisticated German film industry and pushed aside the overt didacticism of von Frisch's earlier films. By the 1940s, filmmakers (including those involved directly in the Reich's propaganda efforts) had to contend with discriminating audiences that hungered for glamorous entertainment rather than overt propaganda. In nature films, too, messages about animals were thought to come across more eloquently when the animals seemed to "speak" for themselves. In von Frisch's films for the RWU, too, the development of the bee "Volk" and its queen, as well as the killing of the drones, unfolds as the ineluctable workings of the hive.

# State of Grace

⚜

On June 22, 1941, Hitler launched his fateful Operation Barbarossa against the Soviet Union. It was the largest deployment of his career—175 divisions comprising over 3.9 million troops. Joseph Stalin had never fully trusted the antiaggression pact with Germany. Nonetheless, he was caught unprepared for the onslaught. Initially, the strike looked promising for Germany, and by November, Hitler's men pressed into Stalingrad. To capture the city would mean to open the gateway to the oil-rich Caucasus and to supply the Reich with much-needed Russian grain. But as is well known, the battle soon turned sour for the Germans, and by November of 1942, Russian troops managed to cut off the food supply to German troops in Stalingrad.

Winter hit hard that year. For days, temperatures hovered at −30 degrees Celsius. By early 1943, hunger, cold, exhaustion, and relentless Soviet fire had whittled down Germany's sixth army from a staggering 500,000 men to 80,000. Despite the troops' ever-growing desperation, Hitler refused them even the slightest retreat. Soviet troops steadily surrounded them and on February 2, 1943, captured the German army, with Friedrich Paulus going down in history with the inglorious distinction of being the first German field marshal captured in battle. According to conservative estimates, close to one million Germans and Russians died in the battle. Civilian casualties are thought to have been even higher.[1]

Not even the relentless German propaganda machine could completely whitewash the fiasco. In a long and rousing speech in Berlin, Goebbels referred to the battle as a "tragic stroke of fate" and thundered that "the onslaught of the steppe against our venerable continent broke loose this winter with a force that surpassed all human and historical imagination."[2] He urged the Volk to rise up in total war. But while Germany would still be able to unleash sporadic bursts of offensive power, the blitzkrieg had come to an end, and triumph would be hard-won if not impossible. German morale on the home front was at an all-time low that winter. Food rations had been reduced again, and workers were expected to put in longer hours for less pay in munitions and other factories that sustained the war. By February of 1943, the Nazis had ordered the closing of bars and restaurants they considered decadent as well as stores that carried luxury items. They also shut down small and nonessential businesses, as all raw materials were now to be funneled into the war effort.[3] The mood in the Reich was grim and about to turn worse as bombs pounded its cities with ever-greater intensity.

The situation in Munich became especially desperate, as the Americans now joined the British in their bombing strikes over southern Germany. The Americans flew their missions from Italy, which had recently been conquered by the Allies. Over the course of three days in mid-July 1944, the US Eighth Air Force flew over 3,300 bombers plus support aircrafts over Munich and dropped close to 8,000 tons of bombs onto the city.[4]

The planes had flown during the day but were forced to carry out their strikes by the wits of their instruments, as the skies were thick with clouds. Although the Germans pelted them with heavy antiaircraft fire from the ground, they encountered no aerial opposition on two of their missions.[5] Alan A. Arlin, captain of the 398th noted in his diary that despite the cloud cover, "Munich is a nice large target." After the third day, he reported, "From what little we could see, it looked like our bombs have not been wasted."[6] Those residents who had thought their city too pretty to be bombed were sorely mistaken. And their city had ceased to be pretty.

In addition to dealing the Nazis a symbolic blow by crippling the Capital of the Movement, the three-punch assault also damaged the heart of the Nazi transportation system. As the *New York Times* noted, "Munich sits astride the most direct route from southeastern Europe to France, the Low Countries and the industrial heart of Germany in the Ruhr."[7] The bombers were debriefed about hitting the BMW plant on the first raid and an aircraft engine plant on the two subsequent raids. But at least one pilot voiced skepticism in his diary about the alleged goals of the mission: "I think these Munich raids are only terror raids."[8]

On July 12, 1944, in the middle of the Allied bombing of Munich, Karl von Frisch drove onto his street on the outskirts of the city and met a surreal scene. Sofa cushions hung from nearby trees and sharp-smelling smoke billowed from roofless houses as scattered fires smoldered. When he arrived where his house had stood, an astonishing pile of rubble greeted him. As he stepped from his car, flames leapt from the basement of what had been his home. He had moved his library there to keep it safe from the bombs that might strike the institute in the city center. He now tried to get someone to help him rescue his belongings, but the only people he could see rushed about trying to put out their own fires. All he could do was try to catch the scattered papers as they fluttered from the conflagration. They were charred and ringed with flames. As he tried to chase down the flaming flakes, his frantic mind worried that among these were private letters he had written to his wife years ago when they had first fallen in love. But the wind blew hot and heedless.[9]

The following day, the institute built just a decade earlier with Rockefeller funds was also hit around lunchtime. Most of the top stories were obliterated. The basement remained intact, and through a stroke of good fortune, none of von Frisch's lab members were hurt.

Over the following month, von Frisch would complete a move to Brunnwinkl. He had already relocated a fair amount of equipment and established laboratories in two of the summer houses the previous summer. But he had continued to commute back and forth to hold classes in Munich and see to the running of the institute. Now there was little question about continuing any kind of operation at the university; the damage was too devastating to hold classes or maintain any semblance of research. He decamped to Brunnwinkl together with the majority of his lab members, including two beekeeping experts, technical assistants, and two cleaning ladies, who had nowhere else to go.[10]

## In Brunnwinkl

It was a perfect day in October in the later years of the war. The smooth lake reflected the mountains and brilliant foliage against a crisp fall sky. A boat glided across the lake with an arch of flowers and greenery reaching over the smiling couple. Karl and Margarete's eldest daughter, Hannerl, sat next to her groom, Theodor. She wore a white gown that reached to her ankles, and Theodor wore his military uniform and cap. Behind them, Hannerl's younger brother, Otto, cut the water with his single oar. Their uncle Otto stood opposite the boy and plunged his oar in rhythmic opposition. The rowers wore matching white shirts and dark vests, and a jaunty brush sprung from the brim of the elder Otto's hat. The boat carried the bridal party from the church in St. Gilgen across Lake Wolfgang to Brunnwinkl to celebrate the couple as husband and wife.

Theodor had joined von Frisch's lab some months earlier after he was furloughed from active military duty. Upon arriving at the Zoological Institute in Munich, Theodor met the boss's daughter Hannerl. And the rest was proverbial history.

Theodor was not the only one to join von Frisch's lab after serving at the front. Martin Lindauer, a young man from a small town in the foothills of the Bavarian Alps, convalesced in Munich from his war injuries. His doctor recommended that he sit in on a class by Professor von Frisch at the nearby university. And so it was that Lindauer stumbled upon a lecture on cell division. Von Frisch spoke with characteristic clarity and showed his students films and beautiful charts of cellular processes. So taken was Lindauer by this refuge of calm and intellectual exploration that he resolved to study biology with von Frisch. Born one of fifteen children to a poor farming family, his dream had been to become a doctor. But he lost function in one arm due to the injuries he sustained in the war and was no longer able to perform many of the manual tasks

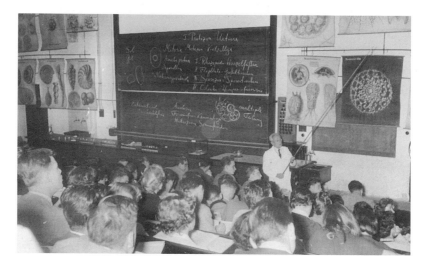

**FIGURE 5.1.** Von Frisch giving a lecture to students on invertebrate zoology. (Nachlaß Karl von Frisch, Bayerische Staatsbibliothek, Munich, ANA 540.)

that would have been required of him as a physician. Despite the setback, Lindauer did fine. Von Frisch soon noticed that the newcomer was an exceptional student.[11]

Indeed, since the early days of his research with bees, von Frisch had a knack for pulling those around him into the orbit of his work. As a young man, he had recruited his famous uncle Sigmund to help out as an observer and thought partner in his experiments. On another occasion, he'd enlisted the help of his oldest brother, Hans, during his investigation of the bees' ability to find food sources over long distances. The goal was to signal a bee's departure from the feeding dish by a relay of sound. Von Frisch would let his brother know by blowing a cattle horn when the first marked bee took flight. Hans, in turn, was to ring a cowbell to let another, yet more distant participant know that it was time to pay attention. As he sat alone on a hilltop listening for the signal, Hans reached into his pocket for a smoke to break the monotony of the wait. With a sinking feeling, he realized he had only remembered to bring his pipe but not the tobacco. He waited for four long hours until he finally heard the horn.[12]

Around this time, Oswald Bumke, a Munich psychiatrist who had also fled to Brunnwinkl because of the war, happened to spot von Frisch's wife Margarete in a field by the road leading into town. But rather than respond in kind to his friendly overtures, Frau von Frisch unceremoniously waved him away. She continued to pace an area within the field, her expression oddly intent. She carried a watch in her hand and occasionally stopped to jot down notes.

Margarete was in the midst of an experiment to help determine how bees might be guided to food sources through scent. In this particular trial, the question was whether the insects could be induced to increase their visits to a nearby cluster of cabbage thistle flowers. Left to themselves, the bees showed little interest in the plants. Now Margarete walked the perimeter of a thistle patch once per minute and recorded the number of bees in attendance. While she performed her diligent survey, others stationed in town began to feed a hive with sugar water that had been infused with cabbage thistle flowers to get the insects to associate the plant's smell with food. Once results were tallied, it became clear that Margarete had captured a significant increase in bee visitations to the flowers during her absorbed walking and scribbling. A few days later over tea, the von Frisches would report their success to Professor Bumke and explain why she had been so oddly curt at their last meeting.[13]

During the late years of the war, more than ever, there were numerous helpers on hand in Brunnwinkl. Von Frisch and his collaborators worked day in and day out without regard for weekends or holidays. The only regular breaks they took were to tend the vegetable fields that had been enlarged by fusing the small garden plots that adjoined each house at Brunnwinkl; with so many additional people to feed, agricultural work on a larger scale was required. But on rare occasions, when the weather was too perfect to ignore, they would set out on spontaneous hiking or skiing excursions. On all other days (except Margarete's birthday), Karl and his students kept to a strict schedule.

Each morning von Frisch rose at four thirty to begin the day's work. He reviewed his students' late-night analyses from the previous day's experiments and then planned the new day's work in meticulous detail. His students, too, would soon get up to prepare. He demanded that everything be in place by 8:00 a.m. sharp to begin the day's experiments. They would conduct two experiments—one in the morning and one after lunch. After dinner, they spent the evening hours analyzing the day's results.[14] And so it would go, day after day, while the rest of the world raged on in deadly conflict.

Brunnwinkl turned out to be a site of major importance for von Frisch's work. During the summer of 1944, he and his collaborators continued to work on the "Russian method" they had begun to investigate two years prior. Soviet scientists had discovered a way to train bees to increase their crop pollination and honey yields. The Russian study had in fact cited von Frisch's earlier work on the animals' sense of smell and scent training. The Russian source suggested large-scale experiments that would allow researchers to "direct bees to fly to flower types in complete accordance with the desire and needs of farmers."[15] The authorities had the relevant Russian sources translated into German and hoped that von Frisch would be able to apply this knowledge to the food effort.

In his memoirs, von Frisch noted that years before the Russians applied the technique, his own beekeeper, Guido Bamberger, had already anticipated their discovery. Bamberger had brought his hives to an area where flowering conditions were especially favorable. Other beekeepers had also hauled their hives to this particular spot, but Bamberger applied a scent-guiding trick to incite the animals to find the food more quickly. Just as the Russians would later do, he cut several of the flowers in question and doused them with honey and sugar water. He then placed the flowers in front of the hive opening and allowed his bees to feast on the offering. Because the scent of the flowers clung to the bees' bodies, their sisters would fly out in search of the smell. Bamberger reported that his bees were the first to reach their goal, and he could boast a higher honey yield than any of his colleagues.[16]

Now von Frisch sought to test and expand the "Russian" approach. The idea was that if a particular flower was especially rich in nectar, the farmer could hasten and increase the bees' approach to the crop. It would also be beneficial to traveling beekeepers, as Bamberger's finding had shown. Overall, von Frisch deemed the experiments a success—depending on the flowers, he reported an increase in honey yields between 25 and 65 percent.[17]

In these trials, von Frisch targeted agriculturally valuable plants in the hope of increasing their seed production. Since many depended on insects for their propagation, increasing the bees' pollination activity would boost the plants' yields. The effort to increase red clover seeds provided an especially promising example, because the plant served as cattle feed, and farmers often could not grow enough to meet their needs. The problem, von Frisch explained, was that the flowers were more suitable to pollination by bumblebees, as those insects had longer proboscises, and the clover flowers' corolla tubes were just a tad long for honeybees. Although the honeybees could technically reach the nectar, they might forego the effort in favor of more accessible alternatives. Unfortunately, bumblebees were relatively rare compared to honeybees, so farmers might do better if the latter could be enticed to increase their visits. And indeed, by training them to feed on the red clover, von Frisch reported a 40 percent seed increase.[18] He was pleased and felt the experiments held great promise for agriculture. After the war, he would bemoan that nobody seemed to pay much attention to these findings. And yet the work would undeniably have a lasting effect; it was during these experiments that von Frisch would once again come to wonder about precisely how the bees found their way to foods.

In one such experiment, von Frisch's longtime colleague Ruth Beutler planned to train bees to feed at a location far from the hive. Perhaps in response to her concern about the time it would take to lead the bees to this faraway spot, von Frisch offered her a more convenient alternative: instead of slowly moving the scented training dish with the bees to the desired spot, he suggested that Beutler simply train the insects to a particular scent at a nearby dish and then move the dish and scent to the farther location.[19] After a scout bee returned to the hive, she would dance and alert her hive mates to the presence of the food. Because the scout bee carried this smell from the training dish into the hive, her recruits would fan out in search of that smell in ever-greater circles. In the meantime, Beutler could just sit and wait for the bees' arrival at the distant spot that was marked with the same scent as the training dish. Or so the theory went.

But when Beutler followed her boss's advice, the bees did not arrive at the distant location as expected. Von Frisch would later remember the frustrated experiment for having prompted him to reexamine his own assumptions about recruitment by scout bees in ways that would have far-reaching consequences.

The weather during August of 1944 was especially beautiful— warm days with little rain provided ideal conditions for bee experiments. During this time, von Frisch and his helpers investigated whether the bees might not indicate the location of food sources to one another by something other than odor. These experiments yielded some of the most important findings of von Frisch's career. And because they would become so seminal to his revised theory of bee communication, they soon took on the burnished gloss of founding myths through their many retellings. But if we look beyond von Frisch's published accounts, a messier and arguably more interesting picture emerges; we glimpse stretches of uncertainty and genuine puzzlement over the means of bee recruitment, pierced by occasional bursts of insight. Only gradually did speculation and searching give way to the elegant explanations that would earn him and his bees their acclaim.[20]

Early in the morning on August 10, von Frisch fed a mild sugar solution to 10 numbered bees on a table set up 10 meters from the hive. He placed the bowl with the solution on a sheet of blotting paper that had been doused with lavender oil.[21] Since he wanted to focus strictly on these 10 animals and their recruited hive mates, all other bees that appeared at the feeding station during the training period were captured and killed. Then, at 9:30 a.m. sharp, he doubled the sugar concentration. Previous investigations had taught him that, while the bees would continue to feed on lower-concentration solutions, they only danced when the liquid reached a certain threshold of sweetness. This made good sense from an evolutionary perspective—all that dancing (by the scout) and flying out to foods (by recruits) cost energy. And since energy is precious in the natural economy of the hive, gains must exceed expenditures to be worth the bees' efforts; the doubled sugar solution was evidently sweet enough to prompt their dances.[22]

At 10:15 a.m., when the bees had fed at the training site and danced back in the hive for forty-five minutes, the food was removed and the trial officially began. Two helpers were stationed at observation sites at 10 and 150 meters from the hive. These observation sites offered the same lavender scent that had accompanied the training dish but without the sugar solution. The observers captured, recorded, and then killed the bees that alighted at their respective stations, so the animals were tracked but did not return to the hive and confound the recruitment puzzle they were trying to solve. During the time of the trial, 340 bees visited the nearby spot while only 8 showed up at the remote site, suggesting to von Frisch that the bees flew preferentially to a distance that corresponded to where the scout bees had been trained at the beginning of the experiment.

Two days later, von Frisch reversed the experimental setup: now 11 bees were trained to feed at a distant spot 150 meters from the hive. There the animals imbibed a sweet sugar solution. An additional 29 bees that arrived at the 150-meter spot during this time unsolicited were promptly captured and killed. After one hour, the food was removed and the observers again recorded the bee visitors at their respective sites. Von Frisch noted that "despite the proximity and the fact that a slight breeze carried the lavender scent to the hive," the nearby observation plate drew only 29 bees, while 38 bees were counted at the more distant site.[23] This meant that when scouts were trained to feed at a faraway site, their recruits preferred the more distant location, despite there being a scented observation spot closer to the hive. This, in effect, confirmed experimentally what Ruth Beutler had observed as a by-product of her failed training efforts—something was amiss with von Frisch's long-held theory of how bee recruits locate foods.

The following day, von Frisch and his helpers hiked even farther with their bees. When they reached a slightly raised meadow at about 300 meters from the hive, they again set up the table, scented blotting paper, and feeding dish. They trained 11 numbered bees by feeding them the sugar solution. Again, two scented observation plates without food were set up, one at 15 and the other at 300 meters distance. Recruits were captured and recorded as they landed at the dishes. This time, only 8 bees arrived at the nearby spot, while 61 bees were caught at the 300-meter location.

The results from these experiments were compelling. When the scout animals were trained to feed nearby, an overwhelming number of recruits showed up at the closer observation site—340 versus 8 bees in the first instance. Conversely, when scouts were trained at more distant locations, such as 150 meters or even 300 meters from the hive, more bees arrived at the faraway sites. Although the bees showed less dramatic preferences for the more distant sites, the fact that there was a preference at all was significant—it meant that these bees, instead of circling randomly at ever-greater distances from the hive in search of food, as von Frisch had believed they would, must have bypassed the nearby sites in favor of the more remote spots. Under the August 12 experiments, von Frisch jotted into his notebook: "suggests distance communication."[24]

But before definitive conclusions could be drawn, his suspicions demanded more testing—were the bees really capable of communicating information about the location of food sources to recruits inside the hive? Further trials would be required with observation stations placed at different distances and directions from the hive. But Brunnwinkl proved tricky for these setups, as the lake and densely forested hills introduced too many variables beyond the researcher's control.[25] To extract meaningful insights, von Frisch needed greater oversight if not control over the surroundings. On August 16, he and his collaborators headed to a nearby farmstead called Aich that offered them more suitable conditions—wide and open flat spaces surrounded the farm that would allow them to

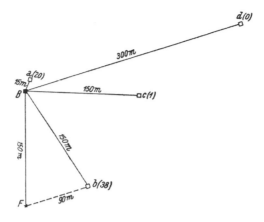

**FIGURE 5.2.** An illustration of an experiment von Frisch performed at Aich near Brunnwinkl on August 16, 1944. Bees were fed at point F during the training phase. Points a, b, c, and d indicate observation stations positioned at 150 and 300 meters from the hive. During the experiment, the most bees (38) appeared at point b, suggesting to von Frisch that the bees' dances communicate information about direction in addition to distance. (Karl von Frisch, "Die Tänze der Bienen," *Österreichische Zoologische Zeitschrift*, 1946, p. 27, fig. 9.)

place their observation and feeding stations wherever their experiments demanded.

Now they trained a few bees with food 150 meters from the hive. But instead of just setting up nearby and faraway observation stations, von Frisch and his helpers arranged different observation spots in a fanlike formation from the hive: one at 15 meters, two stations in different directions at 150 meters, and another at 300 meters. If the scouts simply enticed bees to leave the hive to randomly search for foods by odor, they would reach the closer sites more quickly and tend to congregate there. But if the animals responded to distance information, they would fly 150 meters from the hive to feed because that was where the initial scout had been trained. The setup would also allow von Frisch to discern if the bees indicated direction information to one another.

Over the next days, he and his helpers would repeat variations of this setup with training dishes placed at near and faraway positions and observation stations scattered throughout Aich. These experiments seemed to confirm not only that distance information was somehow conveyed but that the recruited bees also searched preferentially in the direction of the foods that had stimulated the dancing during the training period. In one instance, many bees landed at an observation spot that was 100 meters farther but in the same direction as the training site. Again, von Frisch captured his thoughts alongside his laboratory notes: "The suspicion that the direction of the food source is communicated in the hive increases."[26]

In the meantime, as August drew to a close, the Allies pushed farther into occupied territory and Germany endured additional defeats. While Europe bled, von Frisch buried himself deeper in his bee work. A sunny day was a good day for the Allies to bomb, but it was also a great day for bee work. And for von Frisch, the riddle of the bee "language" was turning into a race against time. Soon the weather would turn to fall and the bees would no longer leave the hive.

By the end of the summer, a lingering question remained: Might the bees emit a kind of aerial scent trail, which other bees then followed to the food source? This explanation was plausible: It was known that bees have what are called Nasonov glands on their hind legs; when the bees evert these glands, they release a powerful odor. As his earlier physiological investigations and, indeed, all of his experiments that relied on scent training demonstrated, honeybees possess a keen sense of smell. In addition, an explanation that relied on odor rather than the insects' ability to communicate held a certain appeal—it was simpler and demanded far less of the animals. And in science, Occam's razor is intended to cut through the thicket of explanation to the simplest possible solution. In animal behavior circles, too, the law of parsimony—known as Morgan's canon—urges practitioners to seek the simplest possible answer to explain a given set of behaviors.[27] And yet despite this utterly sensible stricture, one cannot help but wonder whether von Frisch had been secretly rooting for the more complex explanation of the bees' uncanny abilities to appear where their sisters had previously fed.

To test the possibility that bees attracted one another to foods simply by their own scents, von Frisch planned one more experiment that season. He sealed the scouts' odor glands with shellac. The glue-like substance prevented the insects from emitting their odors but did not diminish their "desire to dance."[28] If the bees still found their way to food, they would have to use something other than odor to communicate location. Conversely, if no bees arrived at the dishes, then von Frisch would have to settle for the much simpler explanation that relied solely on odor. In the latter case, the bees' dances would most likely provide little more than an interesting footnote to honeybee behavior.

But much to his delight, von Frisch discovered that bees recruited by the shellac-altered bees found their way to faraway sites just as readily as they had when their sisters' scent glands had been left intact. Although the issue of odor would haunt him decades later, for the time being, a significant hurdle seemed cleared—the bees conveyed information about the distance and direction of faraway sources to their hive mates by some means other than their own scents. As the summer of 1944 drew to a close, von Frisch could reflect on one of the most productive research periods of his career, and the evidence for some kind of distance and direction indication by the bees seemed as marvelous as it was compelling.

That fall, von Frisch prepared a series of sketches in his notebook that summarized the data from the summer's experiments. In each, he noted the positions of foods and observation sites with respect to the hive, and the number of recruits that visited each location. What is most striking about these compilations is that they offer a map of the terrain in which the scientists and their animals worked as well as of the development of von Frisch's thoughts. And the significance of these drawings lies not just in what was represented but also in what he chose to omit. The pages he kept while performing the trials captured a wealth of information, including barometric pressure, temperature, wind direction, and topographic features. In contrast, these final maps only displayed the pieces of information that he eventually came to consider salient to bee recruitment. Gone were wind and hillsides, while the number of recruits at a given observation station and its location relative to the hive remained. And it was this information that would be presented in the published versions of the experiments. This is not to say that none of the other details made it into published form. But rather than include them in each trial, von Frisch drew on them when he felt the need for additional interpretive resources. For example, when he trained scouts to feed far away but only a few more recruits appeared at the more distant site than the nearby spot, he amplified the significance of this slight difference by mentioning in the text that a slight breeze had carried the scent from a nearby site toward the hive. He explained that since these conditions would bias the results in favor of the more proximate location, the fact that there were more bees at the distant location *at all* spoke all the more loudly in favor of distance communication.

On December 31, 1944, the von Frisches gathered once again to celebrate the holidays in Brunnwinkl. It had been an eventful year. Their home and most of the institute had been destroyed, and nearly all the family members, together with much of von Frisch's staff, had moved to Brunnwinkl to escape further bombings by the Allies and carry on their research. All the children were now there— Hannerl, Maria, Leni, and Otto. Maria still worked in Munich at the Chemical Institute, although her stays in Brunnwinkl were getting longer as well. Leni and her husband, Ekkehard Pflüger, had moved in for the foreseeable future. Ekkehard would soon have to rejoin the war. The two had gotten married in Munich just a few months before Hannerl. Their families had known each other since their childhood days in Rostock when the parents became close friends. The youngest, Otto, had also moved to Brunnwinkl after the family home was bombed. He had just turned sixteen and enjoyed his time away from regular classes. Like his father, Otto evinced a passion for science and animals. During his time at Brunnwinkl, he conducted an independent study of the mice in the area that would later serve as his thesis project.

To all but the most fervent, it seemed now a matter of time before Hitler's Reich would be defeated. As they bid 1944 goodbye, the von Frisches hoped the war would be over soon. But who would decide their fate, how many more would die, when would it be over? And to what kind of a world would the von Frisches' first grandchild be born, as Hannerl had just begun to show? Although they did not know it at the time, that New Year's Eve would be the last time they would all be together.

As the war drew to a close, more and more people arrived in Brunnwinkl; many of them had fled the Russians who were moving into Vienna. At first, it was just a few relatives, some of whom had already spent considerable time in Brunnwinkl. But over the weeks, more and more people arrived. Karl's niece Ute Mohr was just seventeen. Her work group had been disbanded, and she had no place else to go. Luckily, she chose to head to Brunnwinkl rather than Soviet-occupied Vienna. A few months later, an old friend of the von Frisches, Mia Jeannée, arrived with her three young children. They had spent a terrible summer in Vienna with very little food and arrived in Brunnwinkl hungry and exhausted. Mia's husband, Heinrich, was still serving in the army. After he was dismissed, he too had no place to go and no idea where his family might be. He decided to travel to Brunnwinkl, where he hoped Karl's brother Otto would take him in. To his delight he found not only Otto but also his wife and children.[29]

But not all Brunnwinkl endings were happy. Leni lost her husband on April 1, 1945. Ekkehard was killed in the final days of conflict after the Russians had taken Vienna. Hannerl's husband, Theodor, had lost his entire family to a bomb the previous winter. Von Frisch's nephew Anton, who was studying to be a botanist, had been marched into Stalingrad and was never heard from again. The boy's older brother Werner had served as a military doctor on the eastern front, where he contracted a liver disease. He survived the war and imprisonment but would die of liver failure just four years later. Another resident of one of the Brunnwinkl homes lost his son Wolfgang on the Italian front.[30]

The final days of the war brought much uncertainty, and the Viennese were not the only ones on the move. The street that led through Brunnwinkl in those last days of conflict was busy with refugees, mostly farmers from Hungary. Exhausted and fearful, they came with horses and wagons filled with their remaining household goods. Some burdened the lake with their secrets by dumping incriminating possessions from its shores on their way through town.[31]

The inhabitants of Brunnwinkl waited in tense anticipation to see who would reach them first—the Russians or the Americans. It was known that the Russians did not treat their vanquished kindly. Rumors circulated that Russian soldiers had unleashed a brutal campaign of terror and revenge in the conquered areas—women were raped, civilians killed, and belongings destroyed and taken.

Then around midday on May 6, the Brunnwinklers spotted the first tanks rolling along the road from Salzburg toward the resort town of Bad Ischl. They were relieved—the vehicles belonged to the Americans.

But true to human nature, feelings of relief soon gave way to new worries. With an increasing personnel presence, the American military was in urgent need of housing for its soldiers, and relations between the army and the locals grew tense at times. At one point, a group of soldiers entered and searched the von Frisches' home with drawn guns. A pharmacist had been murdered in a nearby town, and now the Americans searched house-to-house for hidden weapons. Von Frisch anxiously opened box after box as the Americans looked on. Friends and extended family members had stored their belongings with them, and von Frisch had no idea what he would find. Luckily, no guns or other incriminating items came to light, and the moment passed. But those living in Brunnwinkl continued to worry that the Americans might seize their homes to accommodate their ever-growing numbers.

One day in June, it seemed, the moment had come. Margarete saw a jeep stop before the house, and four American officers emerged. She nervously approached them to ask what they wanted. To see her husband, they replied. He was out back working, but she would go fetch him. No, they insisted, they would go out to see him instead. And so they rounded the house to where von Frisch was sitting with his bees under an overhang.

# Picking Up the Pieces in Postwar Germany

The Germany that Arthur Hasler encountered when he arrived in April of 1945 was very different from the country he had come to know and love as a young man almost two decades earlier. As a young Mormon, he had taken three years off from his studies at Brigham Young University to proselytize on behalf of the Church of Latter-Day Saints. He had eagerly imbibed the language and culture of the German lands, and years later, as a professor at the University of Wisconsin–Madison, he would recite German poetry to remind students in his ecology courses of the beauty of the lakes they were studying—Mörike, Heine, and Goethe were his favorites.[1] But now, in 1945, there was little time or cause for poetry.

Hasler had been sent to Germany as a member of the US Strategic Bombing Survey, which was part of FIAT (Field Information Agency; Technical), the central clearinghouse for the Allies' efforts to assess, control, and appropriate Germany's technical and scientific knowledge and equipment. Special task forces had been sent even before the war concluded (at times ahead of military lines) to suss out German atomic capabilities and to gather information that might help win the war in the Pacific.[2]

Although Hasler surely had heard and read about the conditions in the former enemy nation, words could not have prepared him for the devastation that awaited. Munich was a mess. Mountains of charred rubble towered over burned-out buildings, and although the Allies had in principle sought to spare buildings deemed artistically or historically significant, there was little evidence of these

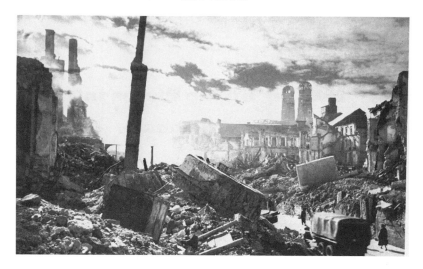

FIGURE 6.1. Munich after the war. The burned-out towers of the Frauenkirche are visible against the right horizon. (Hans Schürer, courtesy Elisabeth Schürer-Necker.)

intentions by the end of the war. Ninety percent of the city's historic core had been obliterated by the bombing campaigns. When the son of the famous writer Thomas Mann returned to his native city, he wrote of the "strange, nightmarish experience" of trying to find his way from Munich's center to his former home.[3]

Hasler made his way through the devastation to Luisenstrasse 14 in the hope of meeting the man whose work he had come to admire. He had been an avid reader of Karl von Frisch's sensory physiological work, especially on fish, because his own research focused on the ecology of lakes. Upon finding his way to the Zoological Institute, he discovered that only the first floor of the building remained. After entering the burned-out shell, he heard from those who had stayed behind that the second and third floors had been blown off during the Allies' bombing raids the previous July. Most of the equipment had been saved in the basement, but the damage was nonetheless severe. Professor von Frisch, they informed him, had moved to the countryside with most of the scientists and students of his lab while his stalwart colleague Ruth Beutler was left in charge of what remained of the institute.

Armed with general directions, Hasler and three colleagues soon set out in search of von Frisch. As they drove from Germany into what formerly was Austria, signs of war eventually gave way to the natural beauty of the Salzkammergut in lower Austria. When they reached the quaint village of St. Gilgen, they asked where they might find the hamlet of Brunnwinkl. Locals pointed out the way, and the men steered their jeep down a narrow road that wound its way along Lake Wolfgang. Awed by the spectacular scenery, the men took in the "angular, wooded mountains" that rimmed the glassy lake and "shot up several hundred feet" along its edge.[4] Finally they reached the small cluster of houses surrounding the old mill. They parked their vehicle and made their way down a narrow path to the mill house. It was here that Margarete met them.

The woman who approached them seemed nervous. How might she help them? They would like to see Professor von Frisch. He was working but she would go fetch him right away. They politely declined. If she could tell them where they might find him, they would be happy to go to him. She pointed them to the back of the house, where von Frisch was working with his bees.

Hasler would later remember the elder scientist intently watching the insects in the observation hive. At the same time, he held a stopwatch to time the animals' movements. From time to time, he took a break from the work and turned to his visitors to explain his research to them.[5]

The previous summer had left tantalizing clues that bees could alert one another to the distance and perhaps direction of food sources. While the evidence that they somehow received such information seemed strong, it was still unclear how exactly this might happen. Most of the focus had been on how many recruits appeared at which observation spots when scouts were trained to food sites. But casual observations of the bees in the hive had already suggested to von Frisch that the speed of the animals' dances changed as the researchers moved the training dishes to ever-greater distances from the hive. At 100 meters, the dances appeared more rapid than those of bees returning from unknown (and presumably more distant) sources.[6] And as von Frisch and his helpers moved food dishes farther from the hive, the dances seemed to slow. Now it was time to more systematically investigate these potential connections.

It was part of this work that Hasler witnessed during his first visit to Brunnwinkl. Von Frisch explained that he would pick a dancing bee at random and count her turns. In his notebook, he noted her identification number, the turns she completed in a fifteen-second interval, and the time of the observation. Then he shifted his focus to another dancer to repeat the counting and recording.[7] At the same time that von Frisch and his visitors sat by the hive, observers at feeding stations noted the numbers and times of the bees that alighted to feed. Later, von Frisch and his collaborators would match each animal observed in the hive with the location at which she had fed just prior, allowing them to calculate the rate of the bees' dances for a given feeding distance.

In addition to the trials explicitly aimed at investigating the relationship between distance and dance speeds, von Frisch was also able to piggyback these observations on other experiments he and his colleagues were conducting at the time. This double-purposed work allowed him to assemble a relatively large number of readings very quickly; in a little over two weeks, he generated 479 observations for distances ranging from 100 to 1,500 meters. And the results were striking—when he plotted the number of turns per fifteen seconds versus the distance of the food sources, a beautiful downward slope emerged. The findings seemed to confirm that foraging distance and speeds of movement were highly correlated; the closer the food, the faster the bees danced.

After the morning's work was completed, Hasler and his companions were asked to stay for lunch. They gladly accepted. After months of shortages, they were served a rare treat—Margarete had prepared fresh vegetables harvested from the family garden. In return, the Americans offered their hosts tins of meat and chocolate, a gesture the von Frisches would remember for decades to come.

Over the next weeks, Hasler was stationed in Salzburg and would continue to visit the von Frisches in Brunnwinkl, spending time with the scientist and his bees and with his family at meals. During this time, a friendship and deep appreciation between the two men took root. Hasler was taken, not just by the man's brilliant work, but also by his "gentle humor, quiet manner and clarity of explanation." He admired von Frisch's determination and grit: "He is continuing his research with what vigor one can muster under limited food intake. His spirit has not been crushed by the Nazis."[8] Through his military connections, Hasler would be able to procure permits for von Frisch to cross the border into Germany. All passages had been sealed after the end of the war, and this intervention allowed von Frisch to resume visits to his institute and family members in Munich.[9]

He continued to explore the problem of bee recruitment, and it was during these experiments—on June 15, 1945—that he made another important observation: He noticed that all the bees returning from a particular food source ran the straight portion of their waggle dances pointing straight downward on the vertically hanging combs.[10] It was after 1:00 p.m. and the food had been set 400 meters northeast of the hive. Von Frisch had on other occasions observed that bees maintained a particular direction when dancing. But the sight of all the bees dancing head-down gave him pause. He noticed that over the course of the day, the dances slowly shifted—by the same angle as the sun moved across the sky, it seemed. In the weeks that followed, von Frisch would fill pages and pages of his notebook with little arrows that captured the direction of the bees' dances with respect to the sun.

To further investigate the precise nature of the relationship between the sun and the angle of the dances, von Frisch and his collaborators returned to the flat open spaces of the nearby farm at Aich. Here they set up feeding stations in the cardinal directions from the hive—due south, east, north, and west, each at 200 meters distance. They observed the bees' dances after they returned to the hive and recorded the angles with respect to the vertical axis.[11]

Gradually a pattern began to emerge. When the food lay northeast of the hive and the sun shone from the southwest—that is, directly opposite each other—the bees directed their dances downward, as the bees had shown so dramatically on June 15. In contrast, when the feeding station was due south of the hive and in line

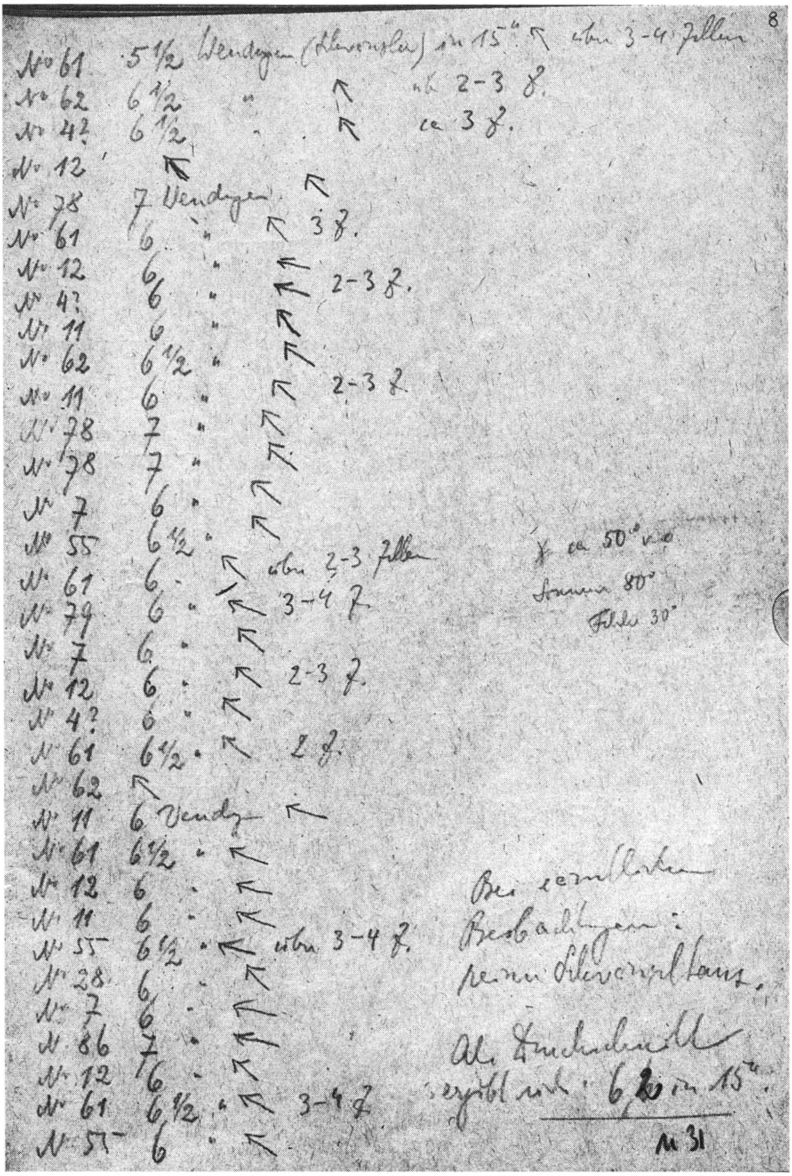

FIGURE 6.2. Von Frisch's lab notes from June 16, 1945. The arrows indicate the direction in which the bees ran the straight portion of their waggle dances. These findings convinced him that the insects indicate the direction of foods in their dances. (Nachlaß Karl von Frisch, Bayerische Staatsbibliothek, Munich, ANA 540.)

with the sun at noon, the bees directed their dances in the hive straight up. When the food lay east with the sun in the south, the bees danced ninety degrees to the left of the vertical, while western foods prompted the animals to direct their dances to the right by ninety degrees. In a remarkable passage, von Frisch translated the bees' actions for his readers: "Dancing directly upward means: you must fly in the direction of the sun to get to the food source. Waggle dance pointing downward means exactly away from the sun is the path to the food. A straight run pointing to the right side suggests, on your flight to the food station, you need to bear right of the sun by the same angle by which the direction of the waggle run deviates from the vertical." His language was uncharacteristically anthropomorphic, but the point emerged crisply: The dances served to communicate, and the bees changed the angle of their movements with respect to the vertical axis of the comb.[12] Gravity, it seemed, stood in for the sun's position during the bee's outgoing flight to foods. And the insects were capable not only of indicating distance but also of encoding precise information about the direction of foods.

Original as these findings were—and they certainly were that—they did not emerge in a vacuum. Von Frisch acknowledged as much in his various publications by citing naturalists who had determined that ants and bees were sensitive to light and in some instances seemed to orient their homing trips according to the sun's position in the sky.[13] In 1914, a physician and amateur entomologist by the name of Rudolf Brun had published a piece on ants that argued that the insects stored and later used a kind of memory imprint (or engram) of the environment that could later be accessed for the purposes of orientation. He and others had also included odors, inclines, topographical landmarks, and local light conditions as part of the insects' navigational tool kit.[14] More recently, Ernst Wolf had published a piece in von Frisch's journal, *Zeitschrift für vergleichende Physiology* (Journal of comparative physiology), that discussed a series of experiments showing that bees relied on the sun to find their way back to the hive. He had worked with the animals in a vast and uniform airfield that offered them little if any useable landmarks. When he kept them trapped in dark boxes for

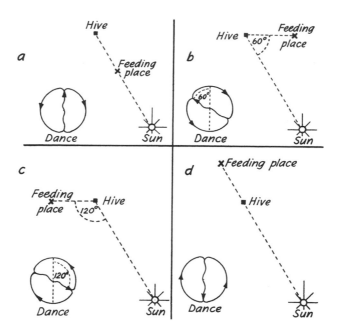

FIGURE 6.3. A schematic showing how the angle of the bee's dance in the hive corresponds to the angle the bee made on her outgoing flight with respect to the sun. Figure *a* shows the bee dancing straight up and down after flying to a food source that lies in a straight line with the sun. (Karl von Frisch, *The Dancing Bees: An Account of the Life and Senses of the Honey Bee*, 2nd ed. [London: Methuen, 1966], p. 135, fig. 87.)

some time, he found that they took flight in a direction that was off by the same number of degrees as the sun had traversed during the bees' captivity.[15] This suggested that the animals remembered the angle they were to fly with respect to the sun and oriented their flights accordingly. These precedents certainly helped von Frisch think about the sun as an important part of bee navigation. But his work was special in that it focused on orientation and communication *between* bees, and the dances became the central means by which that information was conveyed. It was these findings about the bees' waggle dances that had prompted him to quip to Koehler in 1946, "And if you now think I'm crazy, you'd be wrong. But I could certainly understand it."[16]

But alongside the thrill that surrounded the discoveries of this and the previous summer, von Frisch also experienced something decidedly less jubilant. As many of his postwar publications make clear, a sense of regret clouded his findings. Indeed, the new insights contradicted all but the least significant details of his earlier publications on the bee dances, especially his lengthy 1923 work, "On the 'Language' of the Bees." Recall that he had claimed that the round and waggle dances served to communicate nectar and pollen sources, respectively. Now he claimed that the different shapes of the dances had nothing to do with the *type* of food the bees encountered but instead were related to *distance* from the hive. To make matters worse, von Frisch had confidently restated and insisted on his position in response to a 1938 doctoral dissertation. Christian Henkel, a doctoral student at the University of Bonn, had published a dissertation titled "Do the Bees Distinguish Dances?" As the title suggested, Henkel questioned the notion that recruits interpreted the round and waggle dances of the scout bees as meaningful information about nectar and pollen sources. Von Frisch noted how Henkel's "very strongly held assertions" had prompted him to revisit his own experiments on bee recruitment.[17] But at the time, the elder scientist had stood firm and reiterated his position that the round and waggle dances communicated nectar and pollen sources, respectively. Thus, while von Frisch was excited about his new findings, they also forced him into the unenviable position of having to recant his earlier work.

In the first published version of his new results in the Swiss journal *Experientia*, he glossed the mistake lightly. In a footnote, he explained that he had never placed feeding dishes far enough from the hive to observe the bees' waggle dancing after feeding on the nectar-like sugar water: "The mistake was caused by the fact that I would feed sugar water and nectar near the hive, while I generally observed the pollen collectors . . . at more remote food sources."[18] In other words, he only observed the waggle dances in bees that foraged at natural pollen sources at greater distances than he had placed the artificial feeding stations. While this cursory admission sufficed for such a brief paper (which was based on an oral presentation given in Zurich some months earlier), his next publication, a far longer piece published in 1946, demanded elaboration. Here von Frisch again explained that he had not observed bees returning from faraway sources, because he had generally placed the sugar dishes near the hive. But he now called it a "grave error of my former interpretation of the observation."[19] And yet the suggestion that this had been an "error of . . . interpretation" does not fully capture the nature of his misstep. Indeed, it skims past a potentially more serious offense—von Frisch's credibility as an observer. In 1923, he had written about witnessing waggle dances when he fed the bees pollen inside the greenhouse in Munich. But the glass house in which he performed the experiments only measured 27.5 meters.[20] This raised the question of why he had not observed round dances, as the revised theory demanded. The assertion at the time that the bees had run a waggle dance contradicted his new claim that those movements only occurred in bees returning from foods at 100 meters or beyond. So what did von Frisch see in 1923 when he claimed to have witnessed waggle dances inside the glass house? The question clearly vexed von Frisch as well. He described his own oversight as "inexplicable today": "Whether those bees really behaved differently or whether I might have let myself be deceived by a vigorous waggling in a round dance and didn't pay close attention to the path of the dances is probably no longer knowable today." Alas, he conceded, "the latter is more likely."[21] In hindsight, von Frisch's confusion about the waggle dance is perhaps forgivable, as most scientists who study the bees' dances today have abandoned the distinction altogether and instead consider both versions of the dances to be aspects of the waggle dance.[22]

Before returning to the United States in the fall of 1945, Hasler had assured von Frisch that he would try to rally whatever American help he could so that the institute in Munich might get books and articles and hopefully funding to replenish its research budget. He made good on his promise almost immediately after his return to the United States. In letters and published pieces, he brought von Frisch's story to an American readership in the hope that his friend's situation might soon be improved. He asked for donations of scientific literature, clothing, and food, as well as funding. At the heart of these requests was an insistence that von Frisch, rather than having been a contributor to the Nazi scourge, had in fact been another of its victims. In various publications, including a letter to *Science* provocatively entitled "This Is the Enemy," Hasler recounted von Frisch's wartime travails: The scientist's institute had been bombed, as had his home. Not "one page of his library was saved," and his house in Brunnwinkl was looted.[23] "All this," he lamented, "was heaped upon a man who had been oppressed all these years by the Nazis because his grandmother was not 'Aryan.'"[24] Von Frisch, he claimed, had been "a pronounced anti-Nazi from the beginning of the war."[25] And the Nazis only reversed the threat to his job when they realized that he "could be of service to them." Hasler expressed his fervent belief that he and his colleagues could "do science a great favor if we made some gesture to encourage him and assure him of our faith in his type of German."[26]

Though largely correct in its details, Hasler's account of von Frisch's wartime experiences had been subtly tweaked. For example, it failed to mention the very active efforts von Frisch had made to refocus and pitch his work to Nazi authorities in an effort to save his job. As we have seen, rather than having stumbled upon the fact that "von Frisch could be of service to them," Nazi authorities in the Ministries of Education and of Food and Agriculture had been actively lobbied by and on behalf of von Frisch.

There can be no question, however, that the hardships endured by von Frisch and others whom Hasler met in war-torn Europe deeply moved him. He declared himself unable, "as a Christian," to "agree to the collective guilt policy nor collective punishment scheme that appears to be the future plan."[27] Elsewhere he wrote, "Here are men and women who can still contribute richly to science. Many have carried on in spite of political oppressions—they are hungry, without shelter and without heat for their broken laboratories and homes. Their libraries have been burned and blasted."[28]

Hasler also pleaded von Frisch's case behind the scenes in letters to officials at the Rockefeller Foundation. Recall that the foundation had generously funded von Frisch's research prior to the war and paid for the construction of his state-of-the-art institute in Munich. But now the reception was terse—an official replied to Hasler's inquiry that yes, von Frisch was very much in the collective memory of the foundation. But he declined to speculate on his political past: "I have no idea, of course, whether Professor von Frisch played any part in the disaster which Germany brought to the world. Indeed, I find it quite unnecessary to raise the question." He continued, "In a time of military occupation in Germany . . . it seems to us premature and inappropriate for an institution like the Foundation to consider requests from German sources." He regretted the circumstances that compelled them to adopt this policy but hoped Hasler could "sympathize" with their position.[29]

While the foundation officials were not alone in "find[ing] it quite unnecessary to raise the question," other Americans were very much in the midst of grappling with the problem of assessing guilt and culpability. Indeed, Hasler's efforts to establish von Frisch as someone worthy of American help occurred at precisely the time when the American military government of Germany was attempting to sort good from bad in its efforts to democratize, demilitarize, and denazify Germany.

Denazification had been an official goal of the Allies even before hostilities ended. While meeting in Yalta in February of 1945, the three Allied powers declared it their "inflexible purpose to destroy German militarism and Nazism and to ensure that Germany will never again be able to disturb the peace of the world." As part of this effort, they promised "just and swift punishment" for those responsible for the horrors of the war.[30]

The goal was to remove all former Nazis from positions of political, cultural, or economic influence. Not surprisingly, much of the denazification attention in the occupied zones focused on the universities and schools. Universities, in particular, were seen as hotbeds of Nazi activity. In Munich, the Allies closed the university days after taking over the city. While the goal from the beginning was to eventually rebuild the badly damaged institution, the authorities believed that it would have to be thoroughly cleansed before it could once again be trusted to instruct generations of young adults. The new University of Munich would have to be a partner to American efforts to instill democratic values as a bulwark against the dual threats of fascism and, increasingly, communism.

To assist in denazification, all Germans over the age of eighteen who lived in the American zone were required to fill out a questionnaire called a *Meldebogen*. The survey contained 131 questions, which asked them about their participation in national socialist organizations and clubs.[31]

The military government received over 13.5 million completed forms. Of these, about 3.6 million individuals in the US zone of occupation were deemed liable under the law. The military government unburdened the bureaucracy of about two thirds of these liable cases through special amnesty programs. Of the remaining cases, close to 1 million passed through civilian tribunals known as *Spruchkammern*.[32]

Von Frisch's own case was more obvious than most. His classification under the Nazis as one-quarter Jewish as well as his relative lack of party affiliation or club memberships assured him a smooth transition to the postwar regime. Based on his *Meldebogen*, he did not have to stand trial in the *Spruchkammern*.[33] Nonetheless, while von Frisch was not personally obliged to appear, it is worth pausing for a moment to consider this process, as it sheds much light on the context of German science in the postwar period.

The goal of the *Spruchkammern* was to hear evidence and determine the degree of guilt of those who stood before the tribunals. The Americans believed that it was important for Germans themselves to confront their involvement with Nazism and demonstrate their willingness and ability to rid their society of the pervasive influences of its former political system. The juries therefore consisted of ordinary Germans who had not received any formal judicial training but who were considered free of the Nazi taint. These lay jurors were to hear testimony and consider evidence for and against people's involvement with the regime and then hand down a decision that would slot defendants into one of five categories.

These ranged from "major offenders" (category 1) to the "exonerated" (category 5). Category 2 referred to "activists, militants, and profiteers"; category 3, the "less incriminated"; and category 4, so-called "followers or fellow travelers."[34]

By casting individual guilt on a graduated scale rather than on either side of a dichotomy between Nazis and non-Nazis, authorities sought to come to grips with the many shades of moral gray that were the legacy of the twelve-year regime. As the name *Spruchkammern* (literally, speaking chambers) implied, the determination of guilt or innocence depended largely on testimony. That is, efforts to determine culpability relied not on some administrative party list but on defendant and witness accounts in addition to official documents secured by the courts.

But cleansing all aspects of German life from Nazi influences proved a daunting task—testimony was often difficult to interpret and highly context dependent, which meant that those who had lived through the years of Nazi rule were better able than most Americans to pick up on subtle cultural cues to sift truth from self-serving fiction. Not surprisingly, the process relied heavily on native German speakers, many of whom had fled Germany and then joined the US army to fight Hitler.[35] Even so, troubling paradoxes beset the effort. Some who did not enter the party were more ardent Nazis than those who had. Because they had been vocal supporters of the regime, they had perhaps felt less pressure to officially join than less enthusiastic party members. Others had wanted to join because of strong national socialist convictions but had been denied admission for one reason or another. Such individuals could now point to their "refusal" to join the party as evidence of their anti-Nazi convictions. And as with any process that seeks to abstract and codify complex moral choices, mistakes were made—lesser offenders sometimes were made to pay for their sins, while bigger fish escaped the fry.

Tensions also persisted between the Americans, who consistently pushed for more stringent assessments of guilt, and the Germans, who generally erred on the side of lenience. But even within the American camp, members strayed from agreed-upon

party lines. General George S. Patton, who now served as the head
of the military government of Bavaria, bluntly declared at a press
conference that he had "never seen the necessity of the denazification program." He blurted that "the Nazi thing is just like a
Democrat and Republican election fight," and that he felt it "more
important to get the German economic machine running than to
purge industry of Nazi bosses, so that American taxpayers would
not have to foot the bill for food relief." Patton's remarks were as
offensive as they were quotable, and news outlets quickly broad-
cast them. They were met with American outrage and dealt the
beleaguered program's credibility yet another blow. They also cost
Patton his job.[36]

Then in October of 1947 and again in early 1948, after ongoing
cries for greater leniency by the Germans, the Americans suddenly
let up on denazification. In a stunning reversal, they now pushed
for a rapid conclusion to the effort.[37] To hasten the process, they
urged the *Spruchkammern* to downgrade even those labeled offenders (category 2) to mere followers (category 4). The reasons had
less to do with the shelf life of guilt than with the increasing tensions between the United States and the Soviet Union. As the Cold
War began to heat up, Americans drastically shifted their occupation policy; where earlier efforts had focused on punishing Nazi
offenders, they now increasingly assumed a position of rehabilita-
tion.

By the time the Americans declared denazification formally
concluded in 1949, the *Spruchkammern* had classified only 1,654
Germans as major offenders (category 1) from an initial caseload
of 947,000 individuals. This amounted to a mere 0.2 percent. An
additional 22,122 were classed as category 2 offenders (2.3 percent),
and 106,422 were declared category 3 minor offenders (11 percent).
In the universities, this meant that some former Nazis returned to
their positions and taught students by the late 1940s.[38]

In light of these events, it is perhaps not surprising that clear-
ance through the *Spruchkammern* did not ensure that colleagues
at home and abroad considered someone innocent of Nazi convictions. The reputation of Germans was badly tainted in ways that

resisted simple bureaucratic repair. In contrast to the path of dazzling economic recovery on which West Germany would embark, cultural and social wounds endured and festered. The universities also only slowly repaired their damages.

The Nazi legacy would cast an especially long shadow on the life sciences. While American, French, and British scientists rapidly advanced in biology and especially the avant-guarde field of molecular genetics, German efforts languished despite concerted efforts to restore funding sources.[39] In part, this failure may be ascribed to the considerable talent drain Nazi policies inflicted on the nation's intellectual stores. Jewish scientists were pushed from their jobs, persecuted, and murdered. Those who could fled abroad to countries like Britain, the United States, Turkey, and Switzerland. German science had lost some of its most talented and productive intellectuals, and the loss would leave permanent scars. But the numbers tell only part of the story.

As the historian of biology Ute Deichmann has argued, many capable scientists remained in Germany, and the funding they received was at least comparable to that in other combat nations. Rather, she points to a chilling of relations toward German biologists that persisted into the postwar period. Those who had assumed positions that had been forcibly vacated by their Jewish colleagues would be inhibited in their relations with former and new colleagues abroad. Even if they had not been "active" or especially convinced Nazis, their profiting from the plights of others left a bad taste in the mouths of many American, British, and Dutch colleagues. To flourish, science must be international, and many Germans would be conspicuously left off invitation lists to international meetings and scientific societies after the war.[40]

In addition to political considerations, material constraints continued to affect German science in the postwar period. Von Frisch's situation, though better than most, was nonetheless austere, and the working conditions at the institute in Munich remained poor. The university buildings had been badly damaged, and the clearing of rubble and repairs progressed only slowly. The funding and attitude toward science by the military administration was generally considered less favorable in the American than the British and French jurisdictions.[41] These factors tipped the balance in favor of von Frisch leaving his beloved adopted city of Munich for his native Austria when he received an offer of a professorship at the University of Graz in 1946.

The city of Graz had only suffered minimal bombing damage and was now in the British zone. Because the university had been thoroughly "coordinated" by the Nazis, there were now vacancies as a result of denazification. Although Graz University had not enjoyed as high a status as Munich prior to the war, its zoological institute was now attractive by virtue of its being ready for work. Von Frisch was sixty years old—to his mind, the choice was between spending the rest of his career rebuilding the institute in Munich or being able to work on his bees. This was especially compelling in light of his discoveries over the previous summers. Working in Graz would allow him to move freely between Brunnwinkl and the university, as both were in the British zone. Although it would mean being cut off from his children in the American zone, he decided to move (and his children would sneak across the border to join their parents for summers and holidays in Brunnwinkl).[42]

But despite the advantages Graz offered in the early postwar period, funding for the university remained meager, and the housing shortage in the city was severe. So severe, in fact, that Karl and Margarete spent the first two years in Graz living in his office. The workspace served evening duty as their bedroom, kitchen, and bathroom all rolled into one. Margarete resented the evening intrusions of work and lab members into their living space and bitterly missed her children.[43] But the relocation offered von Frisch the peace for the work he so craved.

Two years after the move to Graz, he published a lengthy two-part piece called "Solved and Unsolved Mysteries of the Bee Language" in the prestigious journal *Die Naturwissenschaften* (The natural sciences).[44] The article reviewed all the work von Frisch and his students had accomplished over the past couple of years. While the first part of the piece reviewed how distance and direction were indicated via the speed and angle of the bees' dances, the second part focused on a different aspect of bee behavior that had held von Frisch's attentions in the years following the war—how the insects danced on horizontal surfaces.

Early on in his work on bees, he noticed that the animals sometimes danced on the narrow ledge outside the hive. In contrast to the vertical combs that served as their dance floor inside the hive, these spaces offered a horizontal platform for their movements. Similarly, when he slowly tilted a vertical comb on which bees were dancing to a horizontal position, he observed that the animals resumed their dances after a brief adjustment in direction. He soon realized that the horizontal dances pointed directly at food sources, and when he turned a horizontal comb "like a turntable," the animals adjusted their dances "just like a compass needle."[45] But the question remained: How did the bees know where a given food sources was all the way back at the hive?

Von Frisch assumed that bees relied on the position of the sun to orient these dances. To better understand the relationship between the sun's position and the bees' horizontal dances, he had a special hut constructed of opaque insulation plates. The structure was just large enough to accommodate a horizontal hive and a seated observer. In the enclosed space and by the light of a single red light bulb, von Frisch observed the animals' dances. The red light (invisible to the insects' eyes, as his work on color from nearly two decades earlier had demonstrated) allowed von Frisch to behold the animals' movements. When no sunlight was visible, the animals erratically changed the orientation of their dances, and no two bees seemed to point in the same direction despite their having fed previously at the same location.

And yet when he turned the red light off and instead introduced the beam of a flashlight, the bees' movements quickly assumed a clear and uniform direction. They now oriented their dances as they would if the sun were shining into the hut from the exact position of the flashlight. And by moving the light source, he could "make them dance in any desired direction."[46] The bees adjusted their dances to maintain the same angle with respect to the light irrespective of the actual location of the sun. This was perhaps not overly surprising given what was already known about the animals' ability to perceive light and orient their dances.

But another observation he made during these investigations was decidedly more puzzling: he noticed that, under certain circumstances, the bees indicated food sources accurately on the horizontal even when the sun was not visible and in the absence of an artificial light source. Indeed, to properly indicate the direction of foods, the animals needed to see only a small sliver of blue sky.

To further investigate the phenomenon, von Frisch installed an oven pipe in the roof of the hut that restricted the bees' view of the sky to only a small circle (about ten degrees of arc). The pipe could be angled to point at different areas of the sky and opened or closed at will. When von Frisch uncovered the stovepipe and pointed its mouth at the blue overhead, he observed that the bees oriented their movements correctly. But when clouds passed over the circle, the dances once again became disoriented.[47]

Von Frisch's findings with the stovepipe became even more fascinating. When he pointed the opening at a mirror that reflected a part of the sky into the hive, the bees oriented their dances in the exact mirror-reversed direction from what would have been expected had there been no mirror. When he added another mirror so the bees were exposed to a reflection of a reflection, their dances once again resumed their proper orientation. He reasoned that there must be some means by which the animals could sense the location of the sun by the sheer blue sky.

Conversations with two physicists at Graz and Freiburg helped him generate a hunch—he wondered whether the bees might be capable of inferring the location of the sun by detecting the polarization of the sky.[48] Light exhibits a wave nature and oscillates perpendicularly to its direction of travel. The light coming from the sun is scattered by the particles in the atmosphere. This means that while it starts out as a mixture of rays oscillating in all directions, the light ends up with more of it vibrating in one direction than in others. This phenomenon is called polarization. Since physicists knew that the sun and atmosphere create a unique topography of polarized light in the sky, the phenomenon seemed to present itself as a strong candidate for the bees' orienting capacities. But when von Frisch published his piece on the "solved and unsolved mysteries," he had no firm empirical evidence for his claim, and he knew it. For the time being, he offered as the final unsolved mystery the tantalizing possibility of the bees' ability to detect polarized light.

That fall, just as von Frisch began to investigate the possibility of the bees' capacity to detect polarized light, another visitor came to Brunnwinkl: the Cambridge University ornithologist William Thorpe. Thorpe had done innovative work on the communicative function of bird song and had learned of von Frisch's latest work from the *British Bulletin of Animal Behaviour*. The journal had run an English translation of von Frisch's lengthy paper from the previous year.[49] Thorpe was so intrigued by von Frisch's descriptions and findings that he traveled across the Channel to see the man and his bees for himself.

After Thorpe arrived in Brunnwinkl, von Frisch took him to see the insects. The two men stood side by side peering into the hive as von Frisch pointed out the dancing bees. They proceeded to do "'repeats' of certain of the most crucial experiments." Presumably, Thorpe got to witness how the bees varied the speed and orientation of their dances as the bees' feeding dishes were moved to different distances and locations around the hive. In a write-up of his visit in *Nature*, Thorpe explained that "this memorable experience . . . enabled me to resolve to my own satisfaction some of those doubts and difficulties that come to mind on first reading the work, and convinced me of the soundness of the conclusions as a whole." It was the same tone of skepticism, reluctance, and doubt ultimately giving way to conversion that so often peppered von Frisch's own narratives. Thorpe explained, "The zoologist may, indeed, be pardoned, if at first he feels skeptical—in spite of the immense detail and thoroughness of the investigation and even though it comes from one of the most eminent living workers in these fields of study." In fact, "one experimental zoologist expressed himself to the writer as almost 'passionately unwilling' to accept such conclusions." Thorpe wrote, "Probably many others feel the same—for the implications are certainly revolutionary."[50] This sense of skepticism would continue to surround von Frisch's work, and indeed, even today there are those who doubt that the bees' dances serve a communicative function.[51]

But for Thorpe, his own witnessing of the experiments convinced him of the veracity of von Frisch's claims. He emphasized the implications of the work, which he deemed "astonishing," and urged that the honeybee dance language put the nonhuman animals on an entirely new level: "We are forced to ask ourselves whether apart from human faculties, there is anything comparable known in the animal kingdom." The findings, he continued, "require a reconsideration of some of the most fundamental concepts used in our explanations of the behavior of insects and other animals. The day of Loeb and the tropism theory now seem far away indeed."[52] Jacques Loeb, for his part, had presented animals as machinelike creatures that were compelled to move toward and away from simple stimuli, such as heat or light.[53] What Loeb's lowly animals enjoyed in scientific popularity in the early twentieth century, they lacked in agency and higher functions. To Thorpe, bee communication was exciting precisely because it seemed to suggest an opposing vision of animals and their capacities: "I think it may be said that the performance of the worker hive bee is essentially an elementary form of map-making and map-reading, a symbolic activity in which the direction and action of gravity is a symbol of the direction and incidence of the sun's rays."[54] Thorpe's assessment anticipated much of the tone of postwar discussions of von Frisch's discovery of the communicative function of the bee dances.

But in the immediate postwar period, it served another important function—the article offered von Frisch's work increased visibility with Anglophone readers who would come to know of his "astonishing findings" in ever-growing numbers. In the years following the war, von Frisch continued to speak and write about his findings and to try to improve his material conditions. Almost three years after Hasler's first postwar overtures to the Rockefeller Foundation, a representative of the organization met von Frisch on a visit to Austria. It was Robert Havinghurst's second trip to the former Axis nations on behalf of the foundation. An expert in education and the life sciences, he had been tasked with assessing the general conditions of science in postwar Germany and Austria.[55] He met with von Frisch to discuss his work and plans.

No immediate funding commitments would be forthcoming as a result of the meeting. But the fact that Havinghurst sought out von Frisch on his trip was nonetheless promising. Clearly, the foundation once again considered him a viable candidate for funding. The conversation also offered von Frisch the opportunity to explain his situation. Havinghurst quoted him in his diary as having urged haste: "There is a good tradition of scientific work . . . but it was broken by the Nazis, and slowly the bearers of the tradition are dying out. If the tradition is not renewed with younger men very soon, it will die out."[56] The foundation was not yet ready to reinstate funding for projects in Germany or Austria, but it was beginning to consider sponsoring travel by scientists to the United States to further the exchange of ideas across former enemy lines.

Later that year, Donald Griffin submitted precisely such a request on behalf of von Frisch. Griffin was a young professor at Cornell University who had taken some interest in von Frisch's work. He had read about his experiments in Thorpe's piece as well as in the English translation of von Frisch's 1946 paper. In his own work, Griffin was trying to come up with a cognitive approach to animal behavior in response to the behaviorist paradigm that had dominated American studies of animals. Griffin wanted to bring von Frisch to the United States on a lecture tour. As a result, he lobbied the foundation and colleagues throughout the United States to help sponsor von Frisch's travels. Griffin wrote of his publications and coverage in English-language journals as having "aroused widespread interest among zoologists in the United States." He urged, "If von Frisch could visit this country, confer informally with American biologists, and give a few lectures to acquaint a wider audience with his recent findings, it would aid immeasurably in the evaluation of his discoveries and their use in theoretical and applied work here."[57] Another supporter joined Griffin's efforts to sway the foundation, stating that "von Frisch's many friends at Harvard would be delighted to have him visit here again and . . . we look forward with great expectation to the possibility of having a first-hand account of his recent work."[58]

In late 1948, a letter from the Rockefeller Foundation approved von Frisch's travels. For the second time, almost twenty years after his first visit, von Frisch would travel to the United States. This time, Margarete would accompany him on what promised to be the trip of a lifetime.

# Coming to America

By the time their ship arrived in New York, von Frisch was in bad shape. The journey had exhausted him, and he was racked with indigestion. After giving his first lecture at Cornell, he collapsed in bed and spent the next days recuperating. His host, Donald Griffin, worried that the trip was proving too much and gently broached the possibility of his doing fewer lectures than had been originally planned. Despite von Frisch's assurances that he would soon be well enough to resume his itinerary, Griffin worried. So much so that he sent a mimeographed letter to those hosting the visitor on his future stops. In it, he warned them of the scientist's fragile condition. "Unless von Frisch's state of health improves," he cautioned, "group lunches, teas, etc. should be omitted."[1]

Griffin also recommended that hosts break with the usual question-and-answer format that generally followed the more formal part of the presentation. At Cornell, he found that it had worked well to have audience members jot down their questions on index cards that were then collected and presented to the speaker for answer. This cut down on the direct interaction and also circumvented a problem that had presented itself early in von Frisch's journey—his advanced deafness. Griffin wrote to his colleagues that it was extremely difficult to communicate with the European visitor, especially for non-German speakers. He hoped that this might soon improve, as he had arranged a meeting for von Frisch with members of Harvard's psychoacoustic laboratory to have them fit him with a hearing aid. In closing, he advised hosts to leave a great deal of flexibility in their plans and to "let Mrs. von Frisch advise you just how many group meetings she feels it is wise for her husband to attend."[2]

But counter to expectations, von Frisch soon rallied and was able to resume his lectures. And despite Griffin's concerns, his talks captivated audiences; crisp and clear, the master soon guided his fellow travelers skillfully through the mysterious terrain of bee sensory physiology and communication.

Von Frisch delivered three lectures at the various universities where he was scheduled to speak. Each talk explored an aspect of his work on bees.[3] The first and second were general introductions to the animals' visual and chemical senses, and the third was on his new findings about the dance "language." He began the first talk by explaining the bees' color vision. He recounted his by then classic conditioning experiments in which he trained bees to feed on colored papers to test whether the animals would be able to discern colors. He explained to audiences that his experiments showed the bees' range of color vision compared to that of humans: although the insects did not perceive red, they were able to see yellow, blue-green, and blue, as well as ultraviolet light. He went on to relate the animals' sense of colors to the beekeepers' practice of painting the entrances to hives that housed multiple colonies in different colors. The colors were to guide the animals returning from their foraging flights to enter the correct entrance and thereby avoid being killed by the bees that guard the other colonies. Von Frisch advised that if they wished their efforts to pay off, beekeepers ought to consider the bees' color sense and avoid the practice of marking hive openings with human-centric colors. He concluded the talk by relating the significance of the bees' color vision to special markings on flowers called sap spots—these served to guide the animals to the plants' nectar stores and had coevolved with the insects' sense of color and pollination activities.

The second of von Frisch's talks was entitled "Chemical Senses" and focused on bees' senses of smell and taste. Here von Frisch introduced readers to his experiments with the ceramic scent boxes. The experimenter would drip scented oils onto a small ledge in the box's interior and could later count the number of bees that entered the box. He recounted how he had been able to determine the range and acuity of the animals' sense of smell by setting up an array of boxes, most of which were unscented, alongside one or two that contained scents to which the bees had been trained. He recalled for listeners that the bees' olfactory abilities were comparable to those of humans, although the insects were slightly better at picking out a particular scent from an admixture of odors.

He also recounted experiments in which he had cut off the bees' antennae to determine whether their senses of taste and smell would suffer. Showing fascinating illustrations, he explained how the senses of smell and touch were colocated on these organs and that the animals inhabited a kind of three-dimensional scentscape: "A round scented object may give quite a different sensation to a bee than an angular one." He quoted the Swiss entomologist August Forel, who had "many years ago stated that bees might 'smell' the form of an object as a result of this close relationship between the receptor organs of touch and smell on the antennas." In this talk, von Frisch again reached back to his earliest work to explain how the sensory physiology of the animals had coevolved with flowers. He now also recounted experiments in which he found that the sap spots that visually guide the bees to nectar had also evolved scents to aid in the effort. In addition, flowers had adapted to their main pollinators in the type of nectar they offered. Bees prefered sweeter solutions over more dilute ones, because the insects could only convert liquids of high sugar concentrations into honey. He explained that in response to this predilection by the bees, flowers had over time evolved highly concentrated nectars.

He closed the talk by emphasizing that this knowledge of the bees' senses could be used to guide the animals to food crops. These comments were based on his wartime work for the Ministry of Food and Agriculture. In this talk and on later occasions, von Frisch expressed the hope that beekeepers would adopt this scent-guiding method to increase their crop yields.

In the final talk, he presented the highlight of his work—the honeybee dances. He explained how the distance indication depended on the speed with which the animals ran their waggle dances. To stress this point, he presented his audience with an elegant curve that related the number of turns to the distance of food sources to show a near-perfect correlation. The curve was based on an impressive 3,885 observations. However, while the correlation proved compelling, it was imperfect, and he explained the factors that caused the deviations. First, bees showed considerable inter-individual differences. Second, even between different colonies, there were slight variations in how the dance speeds correlated to the distances of foods. Later this aspect would be confirmed with even more striking results between different bee strains, suggesting to bee researchers something like dialects.[4] Finally, von Frisch reported that winds could cause variations in how the bees' dances correlated to distance. A strong head or tailwind would affect how the animals perceived the distance they had flown.

Given the majestic mountains that surround the Wolfgangsee, it was perhaps not surprising to his audience that von Frisch wondered how bees might communicate the location of a food source that lay behind a large obstacle. In such a case, the insects would be unable to fly to the food in a straight line. But how, he wondered, would they communicate this information to their hive mates—by dancing the circuitous route they had flown or by indicating the direct line to the food's location regardless of the obstacle?

A black-and-white image of students hiking with backpacks around a mountain to perform the experiments accompanied this part of the talk. They had hiked up the Schafberg with the bees to train them to feed on its ridge. Listeners now saw the scientists and their work deeply embedded in the natural surroundings. Surely this image struck a chord with von Frisch's audience. The idyllic picture was taken at the very moment that the country was locked in deadly conflict. And yet there were no traces of militarism, war, or violence. On the contrary, this representation of the bee work exuded a tranquility that harmonized deeply with its surrounding nature. Here and elsewhere, von Frisch's explanations were suffused with a kind of hyperrealism rich with details of the experimental setups that brought time and place to life. It was as though listeners were actually peering over his shoulders to witness these experiments as they took place.

Von Frisch recounted two experiments in which he investigated the problem of navigating an obstacle. In the first, the bees initially

FIGURE 7.1. Hikers performing an experiment on the
Schafberg. (Nachlaß Karl von Frisch, Bayerische
Staatsbibliothek, Munich, ANA 540.)

circumnavigated the mountain by flying a hairpin course around
it. He wondered whether the bees would indicate to hive mates
the direct route or the way in which they had flown the hairpin. To
his surprise, the animals indicated the beeline path, that is, the
straight line from origin to goal. In another experiment, the bees
were trained to fly around a steep mountain ridge. But here the bees
proved themselves less cooperative than in the previous trial; instead of flying *around* the ridge, the animals flew over it in a straight
line. While this flight path was not expected to present the animals
with any problems with respect to the directional indication—they
had indeed flown a beeline—he wondered about what the distance
indication might look like, since the path up and then down the
mountain ridge added significantly to the overall distance flown.
The bees, it turned out, adjusted their dances to account for the

actual distance flown, rather than the shorter distance that would have been required to fly had the ground been level. Von Frisch took stock: "It is doubtless advantageous in such a situation that the new bees are told not only the absolute direction, but the actual distance of flight along the most feasible route." He added the following note to suggest the depths of his amazement and humility in the face of such marvelous behavior: "Even though I have seen this happen, I am unable to comprehend this ability of the bees."

While his talks were overwhelmingly well received, he also encountered some points of resistance along the way. At Yale University, the next stop on his itinerary, he presented his work to the psychology faculty. During a lively conversation that followed, he found himself tangled up in an exchange with Frank Beach, a member of the Yale faculty. Later, he noted in his diary his dismay over the fact that the comparative psychologist had seemed not to make much of a distinction between instinct and intelligence (*Verstand*). For von Frisch, the distinction was as obvious as it was essential—instinctive behaviors are inborn and automatic. They do not require the animal to understand their ultimate goal and are passed on from one generation to the next much like heritable physical traits. On the other hand, intelligent behaviors rely on an animal's insight into a given situation. They allow it to change its behaviors and deviate from the inborn plan. According to the European approach to behavior, instinctive and intelligent behaviors encompassed all behaviors and complemented each other. In contrast, Americans had banished instincts from much of their scientific work since the early twentieth century and focused instead on learned behaviors. Behaviorists—as adherents to this early twentieth-century American school of behavior were called— focused only on those behaviors that were visible. Underlying this approach was the belief that most behaviors were due to stimulus-response conditioning. The behaviorist John Broadus Watson famously declared that if babies were turned over to his capable hands, he would be able to rear them into anything he desired.[5]

But Beach's apparent lack of comprehension of the difference between intelligent and instinctive behaviors rested not so much on ignorance as it did on a "cultural" misunderstanding between the two scientists.[6] The exchange touched on a sensitive nerve in the American behavioral community. Beach did indeed feel that American comparative psychologists had only a vague grasp of the distinction between insight and instinct. But rather than feeling there was no difference, he believed they needed to do a better job of understanding and distinguishing between *different kinds* of behaviors.

Just a few months later, he would deliver an address to the American Psychological Association's Division of Experimental Psychology that lamented this paucity in understanding. According to Beach, American comparative psychologists had focused on learning in the albino rat to the exclusion of all other organisms and kinds of behaviors. While the total number of articles in the *Journal of Animal Behavior* and its successor, the *Journal of Comparative and Physiological Psychology*, had gone up in the last decades, the diversity of species treated in them had fallen dramatically: "From 1930 until the present more than half of the articles in every volume of the journal are devoted to this one species." Beach struck an ominous tone warning that comparative psychology had narrowed its focus to the detriment of its own relevance. To make his point, he invoked the story of the Pied Piper who with his magic flute "rid Hamelin Town of a plague of rats by luring the pests into the river." He warned: "Now the tables are turned. The rat plays the tune and a large group of human beings follow. . . . Unless they escape the spell that *Rattus norvegicus* is casting on them, Experimentalists are in danger of extinction." He quipped, "Perhaps it would be more appropriate to change the title of our journal to *The Journal of Rat Learning*."[7] Thus, while learned behavior was only poorly understood, anything falling outside its purview was hopelessly neglected, and von Frisch's talks had tapped into the heart of American psychology as it was struggling to come to grips with a waning behaviorist paradigm. But von Frisch left the exchange more puzzled than enlightened.

After a couple of days at Yale, the von Frisches traveled on to Massachusetts, where talks had been scheduled at Harvard University. Von Frisch deemed the biological laboratories there "among the most beautiful" he had "ever seen." He was dazzled by the sheer size of the faculty and wrote admiringly of its members' research.[8]

He was especially taken with the experiments of Carroll Williams on the role of hormones in insect development. For one of his experiments, Williams had surgically removed pieces of outer skin from butterfly pupae and replaced them with small pieces of glass. Through these little windows, one could glimpse the animals' tiny beating hearts. Von Frisch marveled that a patient observer could witness the entire transformation of the animals' internal organs as the pupae metamorphosed into butterflies.

In another experiment, Williams had grafted butterfly pupae together like links in a daisy chain. He had excised the upper part of their brains that contained the cells that produce the hormone responsible for metamorphosis. With these critical cells removed, the animals could be arrested indefinitely in their pupal stage. When the researcher reimplanted the brain segment into any of the chain's links, a cascade of metamorphoses was triggered that rendered each of the animals into a butterfly.[9] Von Frisch acknowledged that these experiments might look like little more than "amusing stunts," but he was quick to add that "the expert is impressed by the originality of methods which have actually led to important discoveries."[10] Although he declined to elaborate on the nature of these "important discoveries," his excitement about them was palpable.

During his time in Cambridge, von Frisch also visited the nearby Polaroid Company. There he met its legendary founder, Edwin Land. Land had been a student at Harvard but dropped out to pursue his inventions with polarized film and lenses.[11] It was for this reason that von Frisch visited him; he was given various foils, which he intended to use to test the bees' ability to detect the polarization of sunlight.

In addition to receiving a range of foils from Land, he was also treated to another experience he would not soon forget. Land took him to a back room where he aimed a small square box at him. A mere moment after pushing a button, a thick piece of paper poked out from a narrow slit at the bottom of the device. Land grabbed it by its edge and pulled it from the camera. And there before their eyes, von Frisch's likeness gradually emerged from the matte gray of the paper. "Fantastic," von Frisch enthused in his diary. "Without having to do another hand movement than to pull out the paper."[12] Von Frisch was not the only one to find delight with the "instant camera." The device had been introduced to shoppers at Boston's Jordan Marsh department store as a novelty item the previous Christmas. To everyone's surprise, the entire supply sold out within the first day.[13]

Soon the von Frisches left Boston for New York, where a packed schedule awaited them. On their first day, a reporter from the *New York Times* stopped to interview von Frisch at the American Museum of Natural History. The following day, readers of the *Times* learned that "the first man in the world to establish such close social relations with the insect world" was in the city and due to give his three-talk series at the museum.[14] The *Times* covered each of the talks. After his final presentation on the dances, the paper reported, "The scientist who pierced the secrets of the language of the bees told here last night how he had done it."[15]

While in New York, the von Frisches also visited the New York Zoological Garden, or Bronx Zoo as it would later be called. Von Frisch had met the zoo's director, Fairfield Osborn, at a banquet dinner at Yale. Osborn had invited the couple to lunch and a special tour of the zoo when they were in town, and now they made their pilgrimage to the Bronx. They were awed by the sheer size of the grounds. Osborn had set them up with a car so they could navigate the zoo more efficiently. But von Frisch's review of the zoo was mixed—he lamented that the hummingbirds cooped up in their cages proved only sad reminders of their brilliant beauty in the wild. Nonetheless, he found delight with other aspects and was especially thrilled to see live platypuses that "were kept underground in strict isolation because one hoped for offspring."[16]

Among the other highlights of von Frisch's time in New York was his visit to the American Museum of Natural History. "What riches," he gushed. He was especially impressed with the authenticity of the exhibits: "Not just the apes are real. Every stone, every plant of the foreground comes from the region in which the displayed gorilla family lived and the scenery melts imperceptibly with the naturalistic and artistically painted background." So lifelike were the displays that young visitors could be overheard asking about how one might "get in to water the plants." While von Frisch probably did not share this worry over the encased plant life, he too was awed by the uncanny exhibits: "Nowhere else have I seen the dead creatures displayed so lifelike."[17]

During his stay in New York, von Frisch also met with Warren Weaver, director of the Rockefeller Foundation, in Radio City Music Hall. In sharp contrast to the curt reply from the foundation that Hasler had received to his 1945 letter on behalf of von Frisch's funding prospects, the foundation now invited the visiting scientist to a luncheon to talk to them about his recent work.

Von Frisch was impressed by the Manhattan location and the speed of the elevators—the fifty-fifth floor was reached in just a few seconds.[18] Soon the men were seated around the conference table for lunch. Warren Weaver took the head of the table, with the division leaders framing von Frisch. Over lunch, the conversation meandered from bird migration to orientation. Weaver had coached von Frisch before the meeting that everyone at the foundation was interested in communication between different peoples. He encouraged him to elaborate on his findings on bee communication. Von Frisch obliged and gave an impromptu presentation on his recent findings about the bee dances.

Back in Weaver's office after lunch, von Frisch asked anxiously whether the Rockefeller Foundation was upset with him for having left Munich for Graz. He worried that the foundation might hold it against him that he left the institute that had been funded in the early 1930s through Rockefeller monies. Absolutely not, Weaver assured him. The foundation had "complete understanding" for the decision. The second question he posed to Weaver was whether the foundation would once again be able to help him. Weaver asked what he needed. Money, he replied with characteristic directness. He estimated that he would need around $5,000 annually. Weaver was cautious and warned that he was not in a position to make any promises but that it would be easier for von Frisch to receive money in Austria than if he had stayed in Germany. In Germany, Weaver lamented, so many were in need of funding. He encouraged von Frisch to make a proposal, and if he were granted the funds, he would have free reign over how to allocate and spend the monies. The initial award would be for three years, but if the proposal was approved, Weaver saw no reason why it wouldn't be extended. The following day, von Frisch put together the requested proposal for funding from the foundation.

After their stay in New York, the von Frisches traveled on to New Jersey to visit Princeton University. While giving his lecture to a packed house, a flash of recognition struck him as he caught sight of Albert Einstein in the audience. With his white mane and smiling eyes, he listened attentively to the presentation on bees. Einstein had become a well-known symbol of German scientific brilliance and the rejection of Nazism. In 1933, he had very publicly protested the rise of Nazism by relinquishing his passport and refusing to return to Germany from a trip abroad. Following his departure, the regime maligned his work as emblematic of Jewish thinking and promoted a hackneyed Aryan physics in its place.[19]

After von Frisch's talk, Einstein invited him to his office at the nearby Institute for Advanced Study. Von Frisch remembered fondly Einstein's sense of humor and later noted unselfconsciously that the man possessed a "good head."[20] It is unlikely that they spent much time talking about their histories with Nazism or Germany. The conversation most likely focused on von Frisch's work and their shared delight with intellectual exploration. Nonetheless, both Einstein and von Frisch provided excellent examples of scientists who could be safely embraced by American audiences after the war, because their science seemed apolitical and, more importantly, both men were known to have been at odds with the German state.

While still at Princeton, von Frisch also made a small excursion to the neighboring town of Plainsboro to visit another wonder of the modern age—a futuristic milking factory known as the Rotolactor. Set on 2,400 acres of rolling farmland in the days when peach groves still dotted the Garden State, the dairy housed some 1,550 head of cattle in state-of-the-art conditions.[21] Where ordinary cows grazed on whatever their surroundings offered, these animals imbibed a scientifically composed diet: each day, they received sixty-one pounds of feed comprising fourteen ingredients, including beet pulp, salt, dried milk, concentrated lactoflavin, and alfalfa hauled in from Colorado, Arizona, and Nevada.

Aside from the cows' exacting diet, the daily milking of the animals showcased the dairy's most spectacular innovation. Von Frisch and other visitors stood on a viewing platform from which they could observe the goings-on through a glass window. The animals proceeded to "step in order and apparently eagerly into their boxes of the gigantic rotating disk that completes a revolution every 10 ½ minutes." Von Frisch noted the "white-clad maidens" that "hose down their udders and rears." Next, a "mechanically triggered stream of warm air blows against the bottom and udder. Then the four-part milking hose is attached by hand and one sees the electrically extracted milk collect in a glass container where it is then mechanically emptied and automatically weighed." He learned that the 1,550 cows were milked in just six hours.[22]

Here and elsewhere on his journey, von Frisch noticed strong links between science and agriculture, the clinical atmosphere (with its "spanking clean" cows), and the high degree of mechanization that intimately tied animals to machines in the United States. He would note similar impressions throughout his journey—from high-volume honey production to greyhounds that were kept solely for racing and betting purposes, the United States featured animals that were tightly harnessed to human needs and desires.

Over the Easter holiday, Karl and Margarete encountered more instances of animals that were aggressively manipulated by their human handlers, this time in the explicit name of science rather than agriculture or entertainment. After a strenuous schedule on the first part of their tour, they now traveled to Florida to enjoy a week's vacation at the Yerkes Primate Station in Orange Park.[23] The primate station was home to a wide variety of experimental work, including electroshock therapy, brain surgeries, and deprivation experiments in which chimps were reared in complete darkness or with their hands immobilized for years to determine whether manual dexterity was inborn or learned.

The primate station was nestled in idyllic surroundings amid "great live oaks and silvery Spanish moss." The road leading to the station "ran through the little Southern town" and by an "assortment of fine ancestral homes and humble shanties, past country-style general stores, and a whistle-stop station," as another visitor from around the same time recalled. "Outside the village, the road dipped through a marshy stretch of turpentine woods. When it rose again, there stood the Laboratory: handsome white buildings set among semi-tropical vegetation, the whole enclosed by a stout electric fence. Out of sight of the road, but within easy hearing, was our journey's end, cage after cage of climbing, pounding, noisy chimpanzees."[24]

The evening the von Frisches arrived, commotion filled the air. Loud cries and excited shrieks kept them up as they tried to fall asleep in the unfamiliar surroundings. The next morning, they learned that a chimpanzee mother had given birth to her baby during the night.[25] By the time they went to see the new arrival, the newborn had been freshly diapered and was crying "exactly like a human baby." The animal had been taken from its mother within moments of its birth and was now fed milk from a bottle by its human caretakers. Von Frisch noted that the chimp's mother "did not at all mind, that is, she shows no signs of sadness—as we can see for ourselves—when the baby is taken from her *immediately*."[26]

Another chimpanzee had been taken from its mother some nineteen months earlier and entrusted to the care of a young couple, Catherine and Keith Hayes. The chimpanzee baby, Viki, had just learned to quietly say "Mama" and resembled a human child in many other respects as well. To determine the extent to which the chimpanzee would adopt humanlike behaviors, the Hayeses intended to rear their charge under the same conditions that they would a child of their own. Catherine had taken the ape child into her constant care when the primate was just three days old. They were especially interested in language acquisition and whether the chimpanzee could be taught to speak if it was exposed to human culture and speech from birth.[27] When the couple took Viki to Karl Lashley's housewarming party with the von Frisches in attendance, she played peekaboo and hid behind her adoptive "mother's" skirt when strangers approached. Yet despite the primate's many resemblances to a human child, von Frisch's private assessment of the research surrounding her rearing and of the Yerkes Primate Station in general was not overly favorable. He confided in his diary that the experiments "do not give the impression of [being of a] very high level."[28]

After their Easter rest in Florida, the couple once again boarded a plane. This time they traveled to Ohio, where von Frisch would give a talk at Ohio State University before going on to Ann Arbor, Michigan. From Michigan, they made their way to Chicago. Von Frisch was especially impressed with the spectacular eight-lane Lake Shore Drive. The freeway wound its way along the shore of the seemingly endless lake with the modern city pressed up against its shoulder. They learned of the movable walls that separated traffic at the time; they could be automatically raised or lowered to accommodate rush hour traffic into and out of the city.

After lecturing at the University of Chicago, Karl and Margarete reunited with Arthur Hasler, their postwar friend from Wisconsin. Hasler had driven from Madison to fetch them. On their way back, they caught up at leisure and enjoyed the midwestern landscape as it zipped past their windows. Later von Frisch declared it "an exercise of inadequate means, were I to attempt to describe the beautiful scenery of this road trip." He was no less taken by their destination, calling it "one of the loveliest university towns." Hasler showed them around Madison and took them to his research station on Lake Mendota, where the fish "nearly swim into the laboratory." Von Frisch was duly impressed with Hasler's working conditions and noted with admiration the "eager researchers" who worked for him.[29]

After a few days in Wisconsin, Hasler again took the von Frisches by car, this time to Minneapolis, where von Frisch would present at the University of Minnesota. The remainder of the trip would go by quickly. After a short stay in Iowa, they once again boarded a plane to fly to the West Coast. They made their way from San Francisco to Los Angeles with various speaking engagements along the way.

Then on May 30, Karl and Margarete boarded a plane in Los Angeles and traveled a final time to New York via a brief layover in Chicago. They sat side by side peering through their small window in rapt amazement: "Although we were often above the ocean of clouds, there were moments where we could look onto sand deserts with scraggy mountains, onto the Grand Canyon, [and onto] the many snow-covered peaks of the Rockies." In addition to oceans of yellow blooms, they also enjoyed a beautiful sunset over Lake Erie. Late at night, they arrived in New York.[30]

The final days of their trip were spent again in Ithaca at Cornell with their host Donald Griffin. Griffin had requested that von Frisch demonstrate some of his recent work on bees. He had arranged for a hive from the Department of Entomology of the New York State College of Agriculture to be made available to them. Von Frisch introduced Griffin to his way of working with the insects, and over the following weeks, Griffin sought to replicate his experiments. Later, in the introduction to von Frisch's published lectures, Griffin confessed "without embarrassment that until I performed these simple experiments myself, I too retained a residue of skepticism." But his reservations were apparently short-lived. After a few weeks, he was convinced. He conceded that "much additional work must be done before the dances and the 'language' of the bees are fully understood," but he assured readers that "the important basic facts . . . appear to be established beyond serious question."[31]

The von Frisches left the United States deeply impressed by the country's abundance as well as the sense of progress and optimism exuded by its people and inscribed into its scenery. The brilliance of America's star must have appeared all the brighter against the dark backdrop of Europe in the throes of postwar reconstruction. American grocery stores displayed an astonishing amount of food "hygienically packaged in clear plastic." The von Frisches observed

shoppers claiming a "wire-basket baby stroller" at the shop entrance and marveled that they could fill it with "whatever the heart
desires without any supervision." And at the home of von Frisch's
former colleague Richard Goldschmidt, they were shown their first
dishwashing machine. The Goldschmidts also told them of the
"cleverest things" that were available in the United States. A machine could be filled in the evening with ground coffee and water
and then set to begin brewing early in the morning. When the coffee
finished dripping into the carafe, the device doubled as an alarm
clock and woke its lucky owners to the smell of fresh coffee. They
visited flourishing laboratories stocked with expensive equipment
and populated by so many professors that, von Frisch confessed,
it "made a German biologist a little dizzy."[32]

In addition to a richness of impressions and memories, the visit
was also useful for giving him the opportunity to forge and deepen
transatlantic connections. Many Americans were still wary of their
former enemies, and von Frisch was one of the first Germans they
would meet face-to-face after the war. Because he was known to
have suffered under the Nazis, this contact was greatly eased. His
science was considered "good" in terms of its intellectual caliber
as well as its political innocence. To hear about the bees and their
senses as well as their beneficial application to agriculture gave
people hope that there was science that was both apolitical and
beneficial to humankind. Von Frisch and his bees could serve as
goodwill ambassadors to a nation that was in the midst of global
realignment against the context of an emerging cold war.

The two-month trip allowed von Frisch to present on his latest work to many of the period's most influential scientists. His
lectures left audiences mesmerized. Donald Griffin, von Frisch's
host in Ithaca, spoke of a "pleasing directness and simplicity" to
von Frisch's presentation of his findings and reported that his "recent work has made a most favorable, perhaps I should say a most
dramatic, impression on biologists wherever he has gone."[33] The
visiting scientist, he noted, did "not present" his listeners "with
vague or mystical speculations but rather with phenomena which,
however astonishing they may be, are nonetheless concrete and

readily observed."[34] The comparative psychologist Robert Yerkes declared that he had "never heard a more nearly perfect lecture" than von Frisch's talk on the bee dances.[35]

In addition to gaining important audiences for his scientific work, he was also able to speak informally about the conditions of his home countries, Germany and Austria. Little survives of the nonscientific conversations that occurred around the fringes of his professional interactions. But we can imagine that in the conversations that did touch on his wartime experiences, he was able to tell of his travails under the Nazis and that he had always disapproved of their ways and been glad to see their demise. Undoubtedly such encounters cemented the tenuous transatlantic bonds, for it made obvious that such individuals as Hasler and Griffin as well as the Jewish émigré geneticist Richard Goldschmidt embraced him. And these connections would prove critical to the reception and continuation of his work in the years to come.

But perhaps the most significant meeting for von Frisch's future took place at the Rockefeller Foundation in New York. The application he completed with urgency after his visit led to a grant that would once again support much of his future research, first at the University of Graz and later again at the University of Munich. Against the conflagration that was the Second World War, the Rockefeller Foundation was especially interested in projects that investigated communication. The foundation was eager to promote research that might be applicable to bridging the chasms that separated nations, cultures or, in this case, species. That evening after the meeting, von Frisch wrote in his diary, "Weaver told me before, everybody is interested in the reciprocal communication between peoples [*Völker*]."[36] And it was in this way that Hasler's characterization of von Frisch as the right "type of German" would come to fruition some four years and countless conversations later.

Von Frisch returned to Austria eager to get to work. It was early June and bee season was just beginning. The previous summer, he had started to perform experiments with a polarization foil that seemed to suggest that bees orient their horizontal dances with respect to the sun by perceiving the sky's polarization patterns. He had performed the experiments with a single sheet of foil a colleague had brought him from an earlier trip to the United States. But for the next steps, he needed more foil. The foils he obtained from Land during his visit to the Polaroid Company would prove essential.

Von Frisch reasoned that if bees were capable of analyzing the polarization patterns of the sky, then something in their eyes must serve this purpose. His earlier studies of compound eyes when he was still a student had introduced him to this area of research. His uncle Sigmund Exner had written a treatise on the compound eye, which would later be reissued in English with an introduction by von Frisch. Von Frisch praised the work for its "international significance," and the fact that it was "reprinted today without any of its basic points having been proved wrong in the intervening years."[37] "Today" was ninety-one years after it first appeared. This meant that now and the previous summers when he focused on the bees' behaviors and the more functional aspects of their physiological capacities, he was keenly aware of the intricate physical structures of their visual apparatus. This holistic perspective on the animal—from its evolutionary history down to the cellular level—had also been honed over the years in his work as editor of the *Zeitschrift für vergleichende Physiologie* (Journal of comparative physiology). Nowhere in his work would this vast breadth of knowledge combined with his experimental ingenuity be more brilliantly on display than in these explorations.

For a bee to detect polarization, he reasoned, there needed to be a physical analyzer in the animal's eye. He began to think about whether there might be a prism-like structure that served this function. Following a suggestion from a colleague who specialized in sensory physiology of vision, he focused his attention on the radially arranged sensory cells that are found in the compound visual structure's single eye (or ommatidia).[38] What made von Frisch especially excited about these cells as potential candidates for the polarized light detection devices was the fact that each ommatidium had eight of these cells arranged radially around a central nerve. To

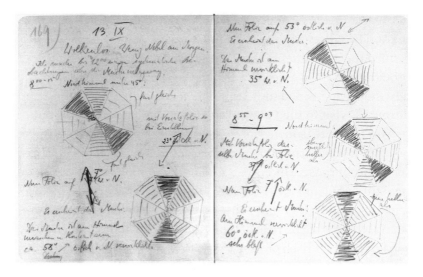

**FIGURE 7.2.** A page of von Frisch's laboratory notebook in which he drew the shading of the different triangular parts of the artificial eye. In this entry, he recorded the time of day, relative cloud cover, and direction in which he pointed the "eye." (Nachlaß Karl von Frisch, Bayerische Staatsbibliothek, Munich, ANA 540.)

test whether such an arrangement would structurally lend itself to an analysis of the sky's light patterns, he decided to build a model eye with Land's foils.

To this end, he cut and radially arranged eight triangles of the polarized foil, like a pizza composed of eight slices. Instead of rounded, the ends of the "slices" were flat and came together as a regular octagon. The axis of polarization of each triangle ran parallel to its base, thereby forming a starlike structure. This was to serve as a model for how the eight rhabdomeres in the bee's simple eye might be arranged about the central nerve. Now when light from the sky was observed through what von Frisch called the "star foil," each of the triangles produced a different degree of light or dark, depending on how much polarized light passed through it. When the device was moved to different parts of the sky, it would cause the eight-triangle pattern to shift.

Imagine polarized light as a standing wave. And now imagine

that a polarization foil acts like a filter with parallel slits that only let light vibrating in that particular plane pass through. Depending on the composition of light in any given part of the sky, the foil will only let through the light that is vibrating in the same direction as the slits. If the foil is pointing at an area with lots of light matching its axis of polarization, then much light will pass through unimpeded making that area of the foil appear bright. In contrast, when the direction of vibration is largely different from the orientation of the slits in the foil, most of that light will be blocked. Consequently, the foil looks darker. Because each area of sky has its own unique composition of polarized light that depends on its location relative to the sun, the eye model gave a unique configuration of triangles in relative light and darkness for each position. Von Frisch related the phenomenon to how an actual bee might see: "Upon regarding the blue sky through this artificial bee's single eye, a characteristic pattern was observed for every part of the sky which is dependent on the sun and therefore changes with the time of day. If the bee is able to perceive the differences in pattern, it can orient itself by a piece of blue sky unambiguously in space according to the position of the sun."[39]

But did the bees actually avail themselves of an analogous polarization filtering process in their eyes to direct their dances or was this merely another interesting way to analyze light? To address this difficult and important question, von Frisch next tested the artificial eye alongside his dancing bees.

He now performed parallel experiments in which he used the star foil alongside actual bees to see if their behavior could be accurately predicted using the foil. For these experiments, he mounted two single sheets of Polaroid foil—one over the artificial, eight-sided eye and the other over a glass window in the cover of a horizontally oriented hive. By noting the direction of the polarization axis on the two sheets and making them adjustable against a circular dial (that told him the exact degrees of rotation of each), he could turn the foils while keeping careful track of their orientation. When the foil over the artificial eye was set at such a position that the eight triangles of the artificial eye remained unchanged, he assumed that the polarization pattern of that portion of sky passed freely through the foil. Now he could adjust the foil over the hive by the same number of degrees—the bees' dances remained unchanged, confirming that the sky's rays passed through the foil unaltered. When he subsequently rotated the two foils by an identical amount and in such a way that it caused a change in the pattern of the artificial eye, the bees also adjusted their dances. After the trial, von Frisch moved the artificial eye, but now without the intervening foil, across the sky until he found the position that gave rise to the same pattern as had the eye with the foil during the experiment. He realized that the bees had "misdirected" their dances by precisely the angle that they would have assumed if the sun had been located in that part of the sky. Thus, the final piece of the puzzle from the previous summer was put in place—von Frisch concluded that the artificial eye offered a reasonable model for how the bees perceived the polarization of the sky when they danced on horizontal surfaces.

A year after he returned from his Strategic Bombing Survey mission in Germany, Arthur Hasler took his family on a trip to his native Utah. He was a lover of the outdoors, and spending time in nature was a favorite activity of the Hasler family. That day, they set out on a hike up Mount Timpanogos. How, he had been asking in the laboratory, could salmon find their way back to their native spawning grounds? As they came to an outcropping of the mountain, Hasler was struck by a sudden wave of recognition. A distinctive smell— from the flowers, earth, and air—washed over him. And then it hit him—it was the smell of home. Here is how he recounted the memory almost four decades later:

> As I hiked along a mountain trail in the Wasatch Range of the Rocky Mountains where I grew up, my reflections about the migratory behavior of salmon were soon interrupted by wonderful scents that I had not smelled since I was a boy. Climbing up toward the Alpine zone on the eastern slope of Mt. Timpanogos, I had approached a

waterfall which was completely obstructed from view by a cliff; yet, when a cool breeze bearing the fragrance of mosses and columbine swept around the rocky abutment, the details of this waterfall and its setting on the face of the mountain suddenly leapt into my mind's eye. In fact, so impressive was this odor that it evoked a flood of memories of boyhood chums and deeds long since vanished from conscious memory.[40]

The solution to the salmon's homing ability, it occurred to him, must somehow relate to their perception and memory of a native smell. Over the following months and years, he set out to prove this hypothesis experimentally, and it would endure as one of the cornerstones of his academic career.

But as is often the case with solutions to recalcitrant scientific problems, smaller realizations tend to accrue and build until they reach critical mass. Hasler himself recognized as much in his later recollections when he credited his insight to the work of two European animal behaviorists which he had read and studied prior to the discovery—Konrad Lorenz's work on imprinting and Karl von Frisch's work on the panic reaction in fish and the scare substance that is emitted when they are injured.[41] It is also likely that von Frisch had talked to him about his work on scent-guiding bees to crops during their time together in Brunnwinkl. Indeed, two years later, Hasler would review von Frisch's book about this work for the journal *Science*.[42] The work discussed not only the power of using odor for navigation but also the bees' abilities to distinguish and remember specific smells.

Thus, Hasler's own work had come to embody the kind of transatlantic exchange of ideas that he had hoped for in the immediate aftermath of World War II. For many, good science by definition meant that it was objective and therefore free of ideology. They believed that aside from the most blatant practitioners of Nazi racial science and the physicians who had been publicly tried in Nuremberg, good scientists should not need denazifying. Proponents of this view, including Hasler, were more open than others to reestablishing international contacts with their German colleagues and firmly advocated for science as an opportunity to build peace and democracy. Von Frisch himself, in the introduction to his English collection of the talks he had delivered in the United States, declared that "scientific work must be international and cannot prosper if confined in a cage."[43] Hasler and others proffered science as a refuge from politics. But linked to this postwar hope that science could promote international peace was a related story that was cocreated during this period—the false notion that scientists had been victimized by the Nazi regime, because it had espoused a strictly antiscientific ideology.

Supporters of von Frisch, such as Hasler and Griffin, were critical to the legitimization and dissemination of the narrative of von Frisch's anti-Nazi comportment. This is not to say that the accounts they offered of von Frisch's experience during WWII were factually inaccurate or deliberately misleading. But they did deploy a good deal of spin. Von Frisch proved an ideal vehicle for shoring up postwar goodwill. He was a bona fide victim of Nazism and yet had managed to continue to do excellent scientific work throughout the war. He then came into the unique position of having conducted work with great freedom in Brunnwinkl with substantial funding from the Nazi government with no taint whatsoever. He was politically tenable when so many others were compromised. Necessary to this construction of character was the subtle shift in how his wartime work was presented. Von Frisch and his supporters actively maneuvered to allow him to continue his research. But these efforts were largely elided in the postwar telling of events. Thus, while German science was in tatters, here was someone about whom the Americans could feel good.

The transformation in von Frisch's standing from an untouchable German from the perspective of the Rockefeller Foundation to a "good German" had facilitated his visit to the United States in 1949. The trip, in turn, enabled him to reestablish funding from the foundation. In June of 1949, the Rockefeller Foundation decided to award $25,000 to von Frisch's institute at the University of Graz: $10,000 was to be used for repairs of the still badly damaged buildings, and three annual installments of $5,000 each were slated

# BEES

THEIR VISION, CHEMICAL
SENSES, AND LANGUAGE

Karl von Frisch

PROFESSOR OF ZOOLOGY IN
THE UNIVERSITY OF MUNICH

CORNELL UNIVERSITY PRESS

ITHACA, NEW YORK, 1950

FIGURE 7.3. The frontispiece to the book of published lectures von Frisch gave during his 1949 visit to the United States. Donald Griffin wrote the introduction and likely facilitated its publication through Cornell University Press, his home institution. (Karl von Frisch, *Bees: Their Vision, Chemical Senses, and Language* [Ithaca, NY: Cornell University Press, 1950].)

to support his research. Indeed, the foundation would generously fund him and his students until the summer of 1964.[44] Thus, the bee researcher had been one of the best-funded scientists before the war and would once again be able to work in highly favorable conditions over the next decades. But in addition to securing generous funds, the trip to the United States also afforded him the opportunity to speak to American audiences about his work at a time when great interest in the European approach to animals was emerging. As behaviorism was on its way out, Americans looked to other approaches that would help them understand animals not merely as stand-ins for humans running mazes in artificial laboratory settings. The European approach brought something new— animals in their natural habitats that were studied for their own sake rather than as mere stand-ins for humans. And the contacts and support von Frisch shored up on this trip would stand him in good stead as he was about to face one of the greatest challenges to his scientific career.

# SEEING BEES

On a hot day in June of 1788, two men stood in a sprawling yard in Switzerland. One looked up, anxiously scanning the sky. He moved about talking quickly while the other turned his body to follow the sound of his companion's voice. Then, as though a tension had lifted from the air, the sky watcher relaxed his attentions. He offered his arm to the other, and the men made their way across the garden to a nearby hive. Shortly after, something changed again—the former sky watcher now focused intently on the small board at the entrance of the hive. The other waited and listened as the watcher turned catcher. He put something—a bee—in a small box or basket he'd fished from his pocket. The men once again spoke quickly as a flush of delight washed over their faces—they had caught the queen.

That singular animal of the hive—the one larger than the others, critical to the survival of the colony, and long celebrated as its sovereign—had posed some of the most stubborn questions to beekeepers and naturalists over the centuries. Was this animal male or female? Aristotle declared, "Some people call the rulers 'mothers,'" but he went on to reject the notion on the grounds that "nature does not give weapons for fighting to any female, and while the drones are stingless all the bees have a sting."[1] To his mind and generations of followers, the case was clear—the singular bee and the colony's workers were male and the drones female. But what of the colony's offspring? Was the king responsible for laying the eggs of the colony? Experienced beekeepers reported that even a colony deprived of the singular bee could produce eggs. And yet those eggs only gave rise to drones, and a colony that did not find a replacement for its sovereign would soon decline and perish. How were the eggs fertilized? Nobody had ever seen bees copulate. Was this a case of parthenogenesis, the controversial theory stipulating that fertilization was unnecessary in certain instances of procreation? Or was there some other phenomenon at play that had escaped even the most determined observer's attention?

The answers to these questions were, quite literally, difficult to see. The animals lived in total darkness and could only be glimpsed through special glass-enclosed hives. But even in these special observation hives, their bodies often obscured what was taking place, especially at the bottom of cells. It was, moreover, nearly impossible to track which animal had done what, as tens of thousands of look-alike workers milled across the combs. And when a single animal was retrieved for closer inspection, its delicate body parts and membranes tore easily, even at the hands of a skilled dissector.

Accurate observations, not just of bees, but of the world at large, lies at the heart of modern science. When Galileo turned his telescope to the heavens in 1609, he argued for a new world order. The moon is not a perfect sphere, as Aristotle's theory had demanded, but instead displays an uneven landscape of mountains, craters, and valleys. That swath of haze known as the Milky Way resolved into myriads of stars through the marvelous scope, suggesting a world that extends far beyond the moon. And precise observations about the phases of Venus and the moons of Jupiter could be marshaled as arguments that planets revolve around the sun and that the earth is not unique with its companion moon.

The telescope and other optical instruments presented a new way of seeing the world, and the revolution in science that occurred in the Latin West was not just an institutional reconfiguration of state, church, and the universities—although it certainly was that—but also a revolution of vision.[2] New instruments for seeing brought together craftsmen (skilled in such arts as lens grinding and clock making), wealthy patrons, and men of science. And in the words of the historian of science Lorraine Daston, "Scientific observers in the seventeenth and eighteenth centuries self-consciously developed novel practices that schooled perception, attention, judgment, and memory." A new generation of scholars believed that truth about nature was to be gleaned not just from books written by the ancients and their seemingly endless trail of commentators but from direct experiences with the natural world. Long-term and increasingly precise observations and calculations coalesced to enable early modern scientists to pluck the earth from its lofty center and toss it into orbit as just another planet destined to traverse its path around the sun.[3]

And it was not just the macrocosm that endured dramatic expansion and revision through this heady cocktail of technical innovation, investigation, and overthrow; naturalists also probed the microcosm beyond previous reaches of the imagination. Under the newly invented microscope, the ordinary—fleas, mold, the tip of a needle—became breathtakingly extraordinary. New and beautiful worlds that had been hiding in plain sight were revealed and lavishly displayed in new books of science.[4]

The world of insects offered a ready fit for this new instrument of the minute. The microscope revealed creatures like the common flea—which to the naked eye appeared as little more than specks— as complex animals with intricate eyes, mouths, and sexual organs. "The microscope lens," argues the historian of science Lisa Jardine, "fuelled scientific speculation about the intricacy of internal anatomies on a tiny scale. It jarred the anatomist's imagination with images of the actual moving intestines and pulsating heart within the transparent flea or louse."[5]

And none was more skillful in the art of microscopy than the Dutch naturalist Jan Swammerdam. His scientific tastes were as promiscuous as his investigations were productive—he discovered red blood cells and valves in the lymphatic system, correctly located muscle contraction in the nervous rather than circulatory system, and established that insects go through their stages of development gradually rather than by abruptly transforming into different animals, as many had believed.[6]

In his *Book of Nature* (*Biblia naturae*), published posthumously in 1737 and 1738, Swammerdam showed the intricate worlds of insects and their insides. Beautiful plates attended the text and laid bare their delicate structures. Although Swammerdam had already in an earlier publication written that the singular bee's "eggs are distributed into two ovaries . . . which are full of eggs, at the time of breeding," it was in this work that he provided virtuoso visual proof that his predecessors had mistaken the bees' sex.[7]

Drones, according to Swammerdam, possess all the trappings of the masculine—two flat testes and requisite plumbing to channel and eject sperm. He also noted that these males have no stinger and are therefore defenseless (*pace* Aristotle). The worker bees he described as "natural eunuchs," although he granted that they "approach nearer to the nature and disposition of the females than the males." Because he was unable to detect ovaries in them, he concluded that they were "like women who having 'lived virgins till they are past child-bearing, serve only the purpose of labor in the economy of the whole body.'"[8]

But the pièce de résistance of Swammerdam's bee discussion was his illustration of the singular bee's reproductive organs. By dissecting countless animals and working closely with artists and engravers, he provided a striking illustration of the creature's sexual organs and heart.[9] The gaze is drawn to the top center of the image where twin ovaries splay out and resemble a pair of miniature wings. Like lavish plumes, the enlarged ovaries, as seen through the microscope, flank and envelop these minute bellows. The picture, Swammerdam explained, is a composite "of parts extracted from two different female Bees." Part *a*, he explained, was derived "from a full-grown impregnated Bee." Part *c* was drawn after "another Bee less perfect, and not as yet impregnated." And from these observations Swammerdamm declared the king female—long live the queen!

As powerful as his observations through the microscope were, the bees continued to prove recalcitrant at the macroscopic level. For five years he kept bees and carefully watched their every interaction and behavior. But despite his best efforts, he was unable to figure out how the animals procreate. "I do not believe," he professed, that "the male bees actually copulate with the females."[10] Instead, he ventured that a fine mist of sperm—an *aura seminalis*—

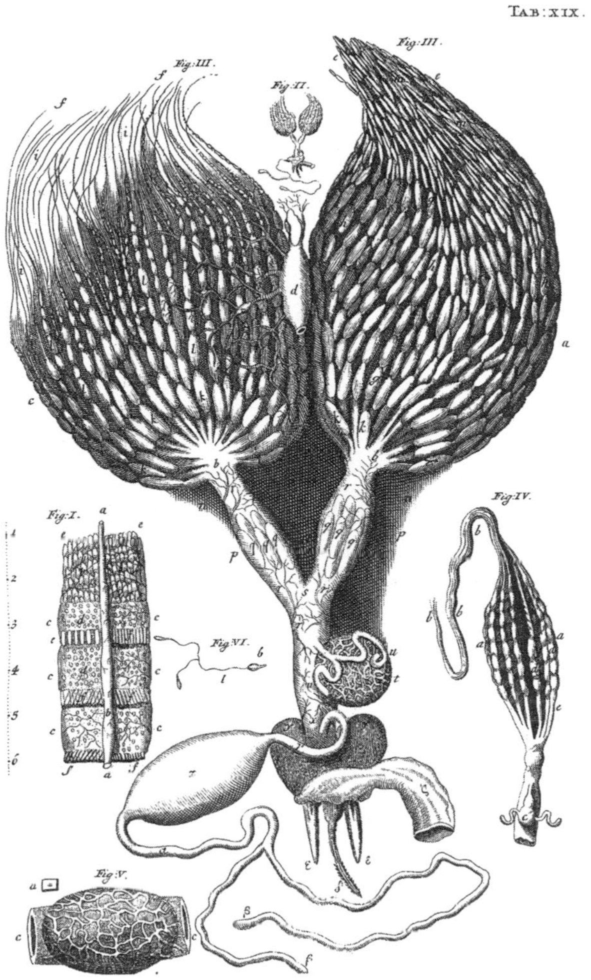

**FIGURE IV.1.** Jan Swammerdam's illustration of the bee's ovaries and reproductive anatomy from his *Biblia naturae* (*Book of Nature*), 1737–1738. (Wellcome Library, London.)

emanated from the drones and seeped into the queen. And so it was that Swammerdam—the man who incontrovertibly declared the queen female by virtue of her ovaries and the drones male as evidenced by their testicles and ejaculatory structures—decided that their communion was nothing more than spermatic mist through the ether. But as odd as this may strike us, his confusion may be excused by the fact that nobody had ever witnessed the queen and her drones in the act. The hive, it seemed, was a sex-free zone.

In 1792, over a half century after Swammerdam's *Book of Nature* announced the sex of the bees with microscopic certainty, another book on bees by the Swiss naturalist François Huber addressed the issue of bee fertility. Huber opened his *Observations upon Bees* with the following disclaimer: "In publishing my observations upon honeybees, I will not conceal the fact that it was not with my own eyes that I made them." Huber's observations were widely considered among the finest of their day. He also became known as the inventor of the "book hive," which induced bees to build single rows of comb onto narrow frames that could be opened or closed like the pages of a book. The hive was so well suited to observation that Huber declared, "There was not a cell . . . in which we could not examine what took place at any moment."[11] But what are we to make of his opening confession that the inventor of the hive that rendered all bee business visible had not observed the bees with his "own eyes"?

As a young boy François Huber was keenly interested in nature and science. Huber's father, John, was a cultured man, a painter, sculptor, musician, and poet. It was said that Voltaire delighted in his company and "valued him for the originality of his conversation." Huber exposed his son François to the arts and sciences in the hope that he might realize his considerable intellectual gifts. But at fifteen, François's health began to falter and his eyesight to deteriorate; soon he was completely blind.

As an adult, François hired a man by the name of Francis Burnens to assist him with his studies and help pass his time in productive contemplation. Though from a humble background, Burnens was bright and literate. These attributes were to give him entrée into the world of science as his employer's eyes. Huber recalled that his new assistant "became extraordinarily interested in all that he read to me." He continued, "I judged readily from [Burnens's] remarks upon our readings, and through the consequences which he knew how to draw, that he was comprehending them as well as

I, and that he was born with the talents of an observer." He took the man under his wing and began to train him as his assistant. When he prompted Burnens to perform experiments in physics, the latter acquitted himself with such "skill and intelligence" that Huber realized he had found his man. He recalled, the "taste which [Burnens] had for the sciences soon became a veritable passion, and I hesitated no longer to vie him my entire confidence, feeling sure to see well when seeing through his eyes."

Soon the two men undertook an intense study of observation—and turned to René de Réaumur, a renowned observer of the late eighteenth century: "I found in this work so beautiful a plan of experiments, observations made with so much art, so wise a logic, that I resolved to study especially this celebrated author to shape my reader and myself at his school, in the difficult art of observing nature." Huber and Burnens formed a kind of intellectual union unlike any other—a complete symbiosis of thought, sight, and hands. The two men worked side by side, one reading, the other asking, noting, and guiding. When words failed them, they passed figures of clay between them to communicate by touch what Burnens had observed with his eyes.

Incidentally, their guide, Réaumur, had written prolifically on insects. Following his lead, Huber and Burnens soon turned their attentions to the mystery of bee procreation. While Swammerdam's microscopic corrections of the insects' proper sexes had found wide acceptance by the time the two men took up their studies, his theory of the *aura seminalis* had gained decidedly less traction. The mystery remained—how was the queen fertilized?

According to Huber, a British observer by the name of John de Braw had reported a small puddle of whitish liquid that pooled at the bottom of the brood cells. This researcher also reported having seen males dip their abdomens into these cells and concluded that the animals sprinkle the eggs with "drops of the prolific masculine fluid." In support of his theory, de Braw proffered experimental evidence as well. On two occasions, he had enclosed a queen and worker bees in an observation vessel. In the first case, he also included a few drones. De Braw reported that only the former arrangement with the drones gave rise to fertilized eggs. In the second case, the queen laid either no eggs at all or ones that remained infertile. Thus, he concluded that bee eggs were fertilized externally, as was the case with fish and frogs.

Huber conceded about de Braw's work that "the experiments upon which it was based appeared well made." And yet doubts remained about external fertilization in the hive: "This explanation appeared specious." Indeed, when he and Burnens cut open the cells that seemed to contain the spermatic fluid, they found them dry. The appearance of liquid had been caused by light hitting the cells rather than the presence of a spermatic deposit. Huber reported another "very strong objection" to de Braw—larvae continued to hatch from eggs long after the males were gone. Each year after the slaughter of the males, the hive was free of drones roughly from September to April. During this time, the queen continued to lay eggs from which bees hatched. Clearly then, it was possible for the queen to lay fertile eggs without these being continuously sprinkled with the "prolific masculine fluid."

Prompted by these conflicting findings, Huber and Burnens designed their own experiments to determine the role of drones in fertilization. Unlike de Braw, they determined that it was quite possible to isolate a queen with only her workers and still get fertile eggs. At the same time, they found that the queen remained infertile, even in the presence of males, if the animals were never allowed to leave the hive. It therefore seemed to them that the queen's fertility was more closely linked to her ability to fly out when the drones took flight than to the presence or absence of males in the hive. Huber concluded that "the females cannot be fecundated in the interior of their hives, and that they must go out in order to receive the approaches of the males." Now all that remained to be accomplished was to watch the animals copulate outside.

And so it came that the two men stood in Huber's sprawling yard in Switzerland on a hot June day in 1788. They had observed males take flight from several nearby hives. The moment was auspicious—these flights by the males generally took place during the warmest part of the day, and the temperature had steadily risen over the course of the morning. And then the queen appeared at the entrance. She had hatched only five days earlier and the men were certain she was a virgin, as Burnens had closely watched her since she'd hatched. "We saw her promenading on the [stand] of the hive for a few instants, brushing her belly with her posterior legs: neither the bees nor the males that emerged from the hive appeared to bestow any attention upon her: at last the young queen took flight." It was then that the men followed her and positioned themselves in the middle of her skyward circles. Burnens squinted into the sky and tried to follow her. The men talked excitedly, then Burnens lost sight of her. He offered Huber his arm and led the blind man back to the hive from which the queen had emerged.

After a few moments, the queen returned, and Burnens caught her at the entrance. She looked no different from when the men had released her. They let her go again, and she took flight. This time, she disappeared for longer than on her previous flight (twenty-seven minutes, according to their watch). Upon her return, Burnens caught her again for inspection. Her body was altered: "We found her then in a very different state from that in which she was after her first excursion: the posterior part of her body was filled with a whitish substance, thick and hard, the interior edges of her vulva were covered with it; the vulva itself was partly open and we could readily see that its interior was filled with the same substance."

Two days after Burnens and Huber observed the queen returning from her flight, they opened the hive and noticed that her abdomen was distended. She had also laid close to a hundred eggs in the cells destined to give rise to worker bees. The men concluded that the "whitish substance" that filled the queen's posterior parts after her nuptial flight must have been spermatic fluid, as it "very much resembled the liquid contained in the seminal vesicles of the males." Upon closer inspection, it would turn out that what appeared as congealed sperm was in fact part of the male's reproductive apparatus that was torn from its body during copulation. Mating the queen is a fatal act.

The altered state of the queen and her obvious fecundity shortly after her flight suggested to Huber and Burnens that the female had come into contact with the male in midair and that he had ejected his sperm into her body. Still, they hoped for firmer proof. Huber concluded the report: "We do not despair of being able, next spring, to secure the last complement of this proof by seizing the female at the very instant of copulation." But despite their best efforts, the "last complement" would remain elusive.

The recalcitrance of the queen's nuptial flight notwithstanding, François Huber has gone down in history as one of the most celebrated bee researchers. Von Frisch, too, recognized Huber as having "made through the eyes of his student the most beautiful discoveries" and reported his pleasure at being able to draw upon "this reliable observer as a witness."[12] While the story of Huber and Burnens is most likely unique in the history of science, there is a way in which the discovery makes an odd kind of sense. Perhaps a blind man was better able to conceive of an event that occurs outside the scientist's visual reach. Science is often fitful when it lacks visual evidence—to see is to believe. But this man was accustomed to imagining and believing beyond the reaches of sight.

For von Frisch, too, problems of knowledge often were tightly bound to problems of vision and visibility. Wearing glasses from an early age to correct for his severe nearsightedness, von Frisch was keenly aware of the significance of sight. His early childhood memories often were of moments spent in quiet contemplation and close study of nature, and careful observation of animals in their natural habitats emerged as one of the cornerstones of the European science of ethology.[13] Not surprisingly perhaps, some of von Frisch's earliest work was on the vision and perception of fish and bees. From their color sense to spatial perception, sight featured as a fundamental sense in his investigations, and his uncle's research on the compound eyes of invertebrates provided an important touchstone for this line of work. Often von Frisch's scientific breakthroughs came on the heels of innovations in visual techniques. The simple system of numbering bees with colored dots revolutionized what could be seen in the hive and beyond; the ability to number and track hundreds of bees had allowed him and his students to follow which bees did what and where. In effect, the technique allowed them to resolve the indiscriminate masses of the hive into individual animals, an essential feature of the science of physiology in which von Frisch had been trained. Von Frisch and his helpers were able to reliably track over decades which animals danced, attended dances, and showed up at observation and food dishes. His student Gustav Rösch adopted the numbering technique to trace generations of bees and determined that the division of labor was tied to the animals' life stages rather than designated at birth, as previous generations of bee theorists had assumed. In other experiments, von Frisch and his helpers hung white sheets along the bees' anticipated flight paths in an effort to track their movements in the air. He was able to determine that they did not lead one another to food sources, which, in turn, opened the possibility that their dances served to recruit hive mates.

Von Frisch was also a pioneer of scientific film and photography. His pieces often featured colored photographs. Entire volumes would contain only black-and-white images, save for the splashes of color in von Frisch's contributions. In his films of the 1920s and 1930s on the senses of animals, scenes of animal conditioning served the hidden function of training audiences to read invisible stimuli—sounds, smells, colors—in black-and-white silent film.

Toward the end of his career, he once again proved himself on the forefront of research by showing how bees perceive the polarization patterns in the sky by creating an artificial bee eye with Polaroid foil from the United States. This work afforded him another glimpse into the vastly complex sensory world of the insects.

But despite these innovations, von Frisch also warned against relying too heavily on instruments to understand the natural world. "When natural science applies overly sharp lenses to her investigations of simple things," he cautioned, "then it can happen that for all her apparatuses, she no longer sees nature."[14] The warning acknowledged that instruments of sight not only bring us closer to the unknown but can also create distance between the observer and the object of observation. These tools of sight are suspected of introducing artifacts that cloud the very vision they are meant to enhance. But on another level, von Frisch's statement also carries an implicit warning of the conceptual apparatus that inevitably mediates our ability to see the world. As noted earlier, he had made a critical error of judgment and observation in his earliest writings about bee communication when he assumed the waggle dances communicate pollen sources. In his regrets about the error, von Frisch allowed for the visual imagination, for all its power, to lead and mislead.

⚜

In the mid-1990s, two bee researchers in Germany, Michael Gries and Nikolaus Koeniger, published a piece in the *Zeitschrift für vergleichende Physiologie* (Journal of comparative physiology), von Frisch's former journal. The researchers described how they had mounted a queen bee carcass on a 2-meter pole. They daubed the queen with one milligram of a synthetically generated substance known as 9-oxo-2-decenoic acid (a substitute for the queen's natural mating pheromone) and affixed the pole horizontally to a mast 10 meters from the ground. They set the rig in motion with an electromotor such that her body spun at 2.5 meters per second, and two Panasonic CCD video cameras filmed the corpse traversing her 12.5-meter orbit. The scientists' sight lines converged on the queen's body at right angles while the cameras captured the position of drones that approached her. Green filters were applied to help distinguish them from the background. Using specially developed software, the scientists then performed single-frame analysis of the recordings to calculate the position of the drones relative to the queen's body. "For each drone on each frame the coordinates of head and tip of abdomen were recorded on data files." The researchers found that "although drones approach the queen dummy from different directions and change their relative position quickly, their body axis is oriented almost directly towards the attracting queen dummy most of the time."[15] From this they concluded that "mating during free flight requires highly developed orientational and maneuvering precision." Although the drones were found to be attracted to the queen pheromone at distances of more than 400 meters, the chase occured visually once the drones were within the vicinity of the queen. And these drones, far from their "gluttonous, fat, lazy, and stupid" forefathers, command the sky with machine-like precision—a characterization that evokes visions of their electronic namesakes deployed in war and commerce.[16]

Much about the biological bees is still unknown. For example, it has not been possible to follow a queen on her flight to track where she goes and what she does. Instead, researchers have opted to control the queen's body close to the ground while observing the drones moving about her. Despite sophisticated technological tools and concerted efforts, issues of visibility persist and vex, and the bees continue to fly beyond our full visual grasp.

# Attack on the Dance Language

⋏

In October of 1973, Karl von Frisch was awoken from a nap.[1] At eighty-six, he was no longer able to keep to the all-day work schedule he'd followed for so many years. Margarete had died nine years ago, and it was some time since he last experimented on his bees. But he still managed to get around and continued to be interested in science. The call was from the Karolinska Institute in Sweden— he had been selected for a share of the 1973 Nobel Prize in Physiology or Medicine. Too frail to travel, he sent his son, Otto, to receive the prestigious award.

That year the prize was also awarded to two of the best-known founders of the discipline of ethology, the European approach to animal behavior studies that brought instincts back into the behavioral scientist's purview. Konrad Lorenz, the charismatic and controversial fellow Viennese whose work on imprinting and aggression had made him a household name well beyond German-speaking Europe, had also been selected as had Nikolaas Tinbergen of the Netherlands. Tinbergen had perhaps done more than anyone else to set ethology on a solid conceptual and institutional footing in the postwar period. After spending much of World War II interned by the Germans for resisting the Nazification of Leyden University, he moved to Oxford. There he built a thriving community of animal behaviorists—his students studied instinctive behavior as well as the ecological relationships of animals to their surroundings.[2]

The Nobel Prize presenter praised the winners for their contributions to science. While he noted that their findings were "based on studies of insects, fishes and birds and might thus seem to be of minor importance for human physiology or medicine," he emphasized that these "discoveries have been a prerequisite for the comprehensive research that is now pursued also on mammals." In particular, he singled out the genetic basis of behavior. Implicitly referencing Lorenz's work on the critical period of imprinting in birds, he pointed to a growing understanding that there were certain events in an animal's maturation that would prove critical to its later development and health. Conversely, these natural developmental processes might be disturbed to grave consequences: "Research into the behavior of monkeys have demonstrated that serious and to a large extent lasting behavior disturbances may be the result when a baby grows up in isolation without contact with its mother and siblings or with adequate substitutes."[3] This came at a time when scientists were intensely focused on the importance of maternal care to the healthy development of infants.[4] Such issues that had previously been studied largely by psychologists were in the postwar period also brought into the domain of biology. The announcer concluded ominously, "The psychosocial situation of an individual cannot be too adverse to its biological equipment without serious consequences. This holds true for all species, also for that which in shameless vanity has baptized itself 'Homo sapiens.'"[5] Thus, the recipients of the prize were recognized, not only for the importance of their work to animal studies, but with explicit emphasis on how their findings might shed light on human development and behavior. And although von Frisch had been more reticent than Tinbergen and especially Lorenz to apply his findings from his animal work to humans, it was precisely such extrapolation by others that had pushed his work into the international limelight by the early 1970s.

But it had not all been smooth sailing for von Frisch and his bees. Indeed, in 1967 two American scientists—Adrian Wenner and his graduate student, Dennis Johnson—published side-by-side pieces in *Science* in which they attacked some of von Frisch's most fundamental claims.[6] A series of exchanges followed, largely in the pages of *Science*, between von Frisch and Wenner and his collaborators.[7] While von Frisch's detractors granted that the bee dances might *contain* information about the distance and direction of food sources, they rejected his assertion that outgoing bees actually *used* this information to locate such supplies. Instead, they argued that more rigorously controlled experiments showed odor alone to be responsible for bee recruitment. Thus, the opening shot was fired in a battle over the bee dances that would occupy the last of von Frisch's scientific energies.

Much was at stake in the debate, and the dispute stirred up some of the most contentious issues facing animal behavior studies. The stakes were especially high in the postwar period, as psychology, linguistics, and animal behavior studies all looked for new ways to interpret animal and human behaviors. Researchers turned away from the restrictive behaviorism that had reduced all behaviors to stimulus-response conditioning.[8] Slowly the old paradigm was giving way to something more complex, more nuanced and probing of the animal's inner life.[9] Onlookers watched the debate over the bee "language" with considerable interest. The American primatologist Stuart Altman declared it "one of the few non-sterile controversies in the study of animal behavior."[10]

The debate did not just turn on the pivot of odor versus dance communication. The two camps were also operating under vastly different paradigms. Although von Frisch and Wenner seemed to meet in the pages of *Science*, where their aims and methods appeared roughly comparable, their visions of animals and their behaviors differed radically. When von Frisch looked into the hive, he saw an animal capable of the most remarkable feat—communication about the precise location of food sources. In contrast, Adrian Wenner saw bees as simple stimulus-response organisms.

Von Frisch's and Wenner's positions developed gradually over decades of research. As we have seen, von Frisch had already begun using the concept of bee communication in the 1920s, although he had a very different explanation of the bee dances then than the one he would later adopt. What is more, well before Wenner became the dance language's archcritic, he too was a supporter of von Frisch's theory. Indeed, his early works, including his dissertation project, were designed to explore aspects of bee dance communication.

But what does von Frisch's use of the dance communication concept tell us of his attitude about animals and their cognitive abilities? In his published writings, he was reserved about such issues. In his *Life of the Honeybee*, he began the section "Mental Abilities of Bees" with a disclaimer: "This section will be short. For on subjects of which one knows little, one should not say much." By the seventh edition, he omitted the section altogether.[11] But in his private writings, he was occasionally more forthcoming than in his published works. In a 1976 letter to Donald Griffin, he reacted to his old friend and colleague's most recent book, *The Question of Animal Awareness*. Griffin, who was establishing the field of cognitive ethology, had raised the possibility in his book that scientists might soon determine experimentally that animals have consciousness. Next to Washoe the chimpanzee, who had been taught American Sign Language, the dancing bees were his most prominent example of why such a cognitive ethology should be possible.[12] Von Frisch responded, "Personally I have never doubted that animals such as apes, dogs, or my bird, with whom I have been close friends for fifteen years, have consciousness and can to a certain extent think." But he continued: "The difficulty here is that one cannot place oneself in the inner life of another being. And we have no idea what the physiological difference between a conscious and an unconscious action [is]."[13]

Thus, von Frisch did not deny that some animals might have consciousness. But he believed there were significant epistemological barriers to proving or exploring its presence scientifically and that the inaccessibility of its inner and outer manifestations dictated scientific agnosticism and restraint. In the case of bees, his skepticism went further still: "I completely agree with you that the bee dances are variable and can be adapted to changing circumstances. But I do not believe that they are voluntary, considered behaviors. One gets the impression that they are an overwhelmingly instinctive performance."[14] For von Frisch, intelligence and consciousness were tightly linked. But the bees' dances—though "variable" and "adaptive"—were indicative of neither. Still, he consistently referred to bee communication as a language, even in the face of criticisms.

Starting in the 1930s, psychologists, linguists, sociologists, and anthropologists commented on von Frisch's work and its relevance to human language.[15] Though most argued that animal and human language differed significantly, none were as chafed by von Frisch's juxtaposition as the Hungarian psychologist Géza Révész. In 1953, Révész published a pointed piece for the *Psychologische Rundschau* (Psychological review) in which he attacked zoologists' inappropriate ascription of the term "language" to animal communication. He warned of the "dangerous anthropomorphism" the misnomer invited. Animal sounds were affective and bore no relation to their meanings. Moreover, their sudden appearance in the animal's development was fundamentally dissimilar to the gradual process of language acquisition in humans. For Révész, these differences pointed to a "great danger"—that is, to deny the uniqueness of human language was to deny human uniqueness altogether. According to Révész, "Language, as such, appears only where living beings are no longer ruled by instincts, affects, and habits, but rather by intentions and goals and by the insight into the suited means [whereby language] can be realized. It is precisely this way of being that counts as the necessary precondition of humanity and separates human life from animal existence."[16]

Von Frisch responded to Révész's piece by emphasizing that the bee dances were both meaningful (*sinnvoll*) and flexible because they varied depending on circumstances.[17] A bee might dance to indicate foods when supplies were scarce but might not for a comparable source in times of plenty. Von Frisch also cited the work of his student Martin Lindauer, which demonstrated that bees did not restrict their dances to food matters.[18]

Lindauer had shown that during the swarming period, when a portion of the colony split off to form a new colony, a sort of negotiation took place. Returning bees waggle danced to indicate the location and quality of potential nesting sites. Based on the dances' information, hive mates would choose and populate the site best suited to their needs. In addition, the waggle dance was used to indicate water sources and other materials necessary to the colony.[19]

Now von Frisch argued that because the waggle dance indicated

FIGURE 8.1. Karl von Frisch (in the lederhosen) with a group of students in the 1950s. On the far left is Martin Lindauer. (Courtesy Rosemarie Lindauer.)

distance and direction in the same way when referring to something other than food (such as nesting sites and water), precisely the sort of relation existed between movements and their meanings that Révész had denied.

Von Frisch also rejected the anthropomorphic label Révész had flung his way by pointing to the discreet caveat he had placed on the term "language" for some thirty years—the quotation marks: "I seek to counter the misconception of [having] an anthropomorphic understanding by generally placing 'language' in quotation marks." These were intended to alert readers that animal and human language were not identical and to signal his own awareness of these differences. But while von Frisch felt the bee "language" should not be equated with human language, he also urged that it not be reduced to what other animals do: "It would be equally wrong and a denial of the facts if one would place the bees' method of communication on par with the warning calls of many other animals or the similarly simple 'communications of social associations.'"[20] According to von Frisch, bees were unique among nonhuman animals in their native use of symbols, and their dances were considered more complex than any other form of animal communication. Thus, in his eyes, the difference between language and "language" was one of degree rather than kind.

In 1967, Wenner and Johnson published their challenges of von Frisch's step and fan experiments in *Science*. As we have seen, von Frisch had originally performed these experiments in 1947 and continued to perform variations of them into the early 1960s to determine the accuracy with which bees use dance information.[21] Recall that he would train bees to feed at a particular feeding station and then set up additional observation stations, either in straight lines or semicircles for the step and fan experiments, respectively. Scout bees trained to feed at the original site returned to their hive mates and danced to indicate just that location. Von Frisch then filled all the dishes with food and counted attending bees to determine whether recruits preferred the sites indicated by their fellows' dances over others.

Wenner and Johnson modified the experiments for their *Science* pieces. In addition to bees trained to feed at only one site, they also trained a lighter strain of bees—distinguishable only by color, not behavior—to feed at all the other dishes of the step and fan arrangements. Their question was whether recruits would follow their hive mates' dances and visit only the original feeding station or whether they would be (mis)led to feed indiscriminately, or even preferentially, by the sight of bees at other locations. Counter to von Frisch's theory, they found that bees visited all additional food dishes, despite their having had the benefit of dances about just one site. They concluded that the mere presence of feeding bees was sufficient to attract other bees to foods. Although they still granted that the bees' dances contained information about food site locations, they argued that something other than the dances had attracted the bees to foods, and deemed odor the most likely candidate.[22]

In his rebuttal to the piece, von Frisch drew on his vast reservoir of bee work that spanned the previous decades.[23] To prove that bees in fact used dance information rather than scents, he recounted the obstacle experiments for which he had trained bees to fly around a tall building or mountainside to foods.[24] Upon their return, dancers indicated the true direction of the food source, not the roundabout path they themselves had flown. Bees that attended these dances flew in a "beeline," rather than the dancers' diverted paths, indicating to him that dance information, rather than scent trails, guided their searches. He also faulted his detractors for having placed their dishes too close to the hive. Whereas Johnson, in his fan experiments, had set dishes no farther than 270 meters from the hive, von Frisch had tested distances of up to 1,250 meters. He now argued that bees could not be attracted to odors at such great distances. He also criticized Wenner's claim that bees' own scents guided recruits. Already, in 1944, he had rejected this notion through the experiment in which he shellacked shut the bees' Nasonov glands.[25]

Wenner and Johnson's reply is revealing. They claimed that their findings did not disprove the "dance-language hypothesis." Rather, in the spirit of parsimony, they wished to "indicate that a more simple interpretation exists for previous experimental results." Odor could just as well account for how bees found foods and was a simpler and, therefore, preferable explanation. They rejected von Frisch's shellac experiment and argued that the substance itself could have contributed to local odors and guided the bees to foods. They also pointed to the difference in foraging success for bees that had and had not already visited the food source. Those bees that already had visited a food site could rapidly locate it again after hive mates carried its scent into the hive. In contrast, those bees that had never visited the food fared more poorly. To Wenner and Johnson, the data suggested a "probabilistic search pattern" rather than a "goal-directed flight." They urged a more holistic approach that did not focus on individual animals' behaviors but instead viewed the hive as an integral part of a larger odor landscape: "We find that the hive, its surrounding environment, and its past history are part of a dynamic system and must be studied as such."[26]

In 1967, Wenner struck once again at von Frisch's theory in a piece for *Physiological Zoölogy*. This time, he teamed up with his later *Anatomy of a Controversy* coauthor, Patrick Wells, and James Rohlf, a statistical biologist at the University of Kansas. In the article, the authors performed sophisticated factor analyses of the bees' dances. They examined parameters such as sugar concentration and viscosity, time of day, and light intensity to determine whether aspects of the dance corresponded to the richness of food sources, as von Frisch had claimed. They found no significant correlation and argued that even if they had, there was no proof that other bees actually *used* this information, as too little was known about "the capabilities or state of the receiver."[27]

The paper offered no new experimental data but was laden with mathematics. The then eighty-one-year-old von Frisch found little into which he could sink his teeth. He once again turned to his trusted mathematical advisor, the myrmecologist Rudolf Jander. For his *Dance Language and Orientation*, von Frisch had already relied on his colleague's mathematical talents. In a letter to Jander, he lamented that one of the corrections he had adopted from Jander made him "feel a bit like flying blind and without instruments," but he also confided that "unfortunately I can't understand [Wenner's] calculations at all."[28] This time von Frisch's appeal was similarly pathetic, although his reference to "a devastating critique of the 'dance language hypothesis'" may be read as a rare instance of Frischian sarcasm: "The result is a devastating critique of the 'dance language hypothesis,' the justifications of which unfortunately remain mysterious to me because of my lack of statistical talent. Could you perhaps have a look at the work and make it understandable to me insofar as this is possible?" He closed dismissively: "In the nonstatistical part he repeats the same rubbish he already offered in *Science*."[29]

Despite von Frisch's efforts to convince his young adversary, the bee language's purported communicative function gradually crumbled before Wenner's eyes. He believed that bees could find food sources by odor alone and that von Frisch's claims did not withstand mathematical scrutiny. He began to write polemical pieces that accused von Frisch and his supporters of romanticism and a blind embrace of the exotic. To Wenner, a deep association with an organism often led to anthropomorphic interpretations: "Many animals have remarkable structures and/or behavioral patterns. The discovery of yet another remarkable event among animals will find a ready acceptance in a basically optimistic audience. . . . One need observe bees for only a very short period of time before becoming impressed with the many human-like actions they have. An anthropomorphic interpretation may soon follow, even if subconsciously."[30] To be sure, this claim by Wenner could be backed up with countless examples over the centuries. As we have seen, the bees tended to accrue the values and mores of those who studied them.

But there was another profound disconnect that prevented the two scientists from reconciling their differences. To Wenner, the assumption that the dances' information *content* necessarily implied information *use* by other bees was hopelessly teleological. Time and again, he railed against what he called an Aristotelian or Darwinian teleology in biology. To his mind, explanations that sought to answer "why?" or "what for?" questions, hampered scientific progress in answering "how?" questions and promoted misleading conclusions. The approach failed to take into account that not all behaviors are adaptive nor may their purpose be inferred readily.[31] In the case of bees, he argued that the dance information "may well be only *a symptom* of what a foraging bee has experienced as it flies between hive and food place, *not a signal* for other bees."[32]

For von Frisch, on the other hand, complex instincts demanded functional explanations. A deep-rooted commitment to evolutionary explanations marked what was perhaps the most unbridgeable difference between them. Wenner's admission of the bees' complex dances, followed by his steadfast denial of their communicative function, amounted to an incomprehensible, biological non sequitur for von Frisch. He expressed to colleagues his exasperation with Wenner and his allies: "What always amazes me is that he finds so many supporters in the United States, even though he maintains that the direction and distance information goes unheeded by the alarmed hive mates, especially since there are no elaborate behaviors without specific functions. How could such a differentiated dance have evolved if it had no biological significance? Apparently Wenner and his followers don't trouble themselves with this."[33] Thus, what amounted to illegitimate teleology to Wenner was binding necessity for von Frisch. Surely the bees did not dance simply for the sake of dancing.

In 1970, *Science* ran another piece on the bee dances by three Caltech undergraduates: James Gould, Michael Henerey, and Michael MacLeod. The authors had modified the fan experiment to address Wenner and Johnson's challenges to von Frisch's work. In addition, they had attempted to control for any odors that might have helped guide bees to their foods. While they granted that "under certain circumstances, odor cues alone might suffice," on balance, they came out in von Frisch's favor: "Under the experimental conditions used, the directional information contained in the dance appears to have been communicated from forager to recruit and subsequently used by the recruit."[34] The article's first author, James Gould, went on to do his graduate work at Rockefeller University under von Frisch's longtime supporter Donald Griffin and continued to publish research in support of the communicative function of the honeybee dances.

Griffin had of course been a longtime supporter of von Frisch and the chief architect of his 1949 lecture tour of the United States. Nearly thirty years after their first letters, he wrote to von Frisch of his "deep admiration for all [your] work over many years not only on the language of the bees but the hearing of fish and so many other topics. . . . My admiration (I am almost tempted to call it 'hero worship') has, as you know, been of very long standing." He also commented on the debate with Wenner and Wells: "I continue to be amazed, even outraged, at the stubborn refusal of Wenner and his colleagues to face the 'fact of life' concerning your brilliant discoveries. I hope, but with no great confidence, that Gould's recent experiments will convince even them." But he regretted that "there seems to be an almost ideological reluctance, bound up I suspect with the strong current of behavioristic reductionism that has been so prominent among behavioral scientists in America."

Griffin linked Wenner's obstinacy to his behaviorist tendencies but felt the situation was not hopeless. He sensed that behaviorism had run its course and was giving way to other approaches: "I am convinced that this [behaviorist] tide has turned, and I can assure you that despite the many foolish publications from Wenner and his colleagues, almost no one takes them very seriously."[35]

Griffin was not alone in viewing the debate in terms of behaviorism. In a letter to his friend in the United States, the evolutionary biologist and modern synthesist Ernst Mayr, von Frisch had inquired whether *Science* might be ill-disposed toward him because of postwar anti-Germanism. Mayr, himself a German, reassured von Frisch that this was unlikely. At the same time, he pointed to what he considered a more probable cause for von Frisch's troubles with the journal: "I do not think that *Science* can be designated as hostile to German scientists. On the other hand, there are indications that several of the members of the editorial staff are extreme behaviorists and reluctant to believe in genetic programming of behavior."[36] Von Frisch himself never emphasized the genetics underlying the bee dances and in private correspondence voiced his disapproval of what he felt to be geneticists' overconfident reach.[37] Still, the letter makes clear that von Frisch's work was seen as evidence against behaviorism and in favor of genetically determined, instinctive behaviors.

Many in the animal behavior community welcomed Gould's work as the final word in the controversy and hailed it as the vindication of von Frisch's theory. In 1979, the animal behaviorist William Thorpe discussed the debate in his history of ethology. "In recent years," he wrote, "there have been disbelievers who thought they had shown that the dances of the bees were not really communicative and that it could all be explained by conditioning to particular odors in a particular environment." He decisively came down on the side of von Frisch: "Suffice it to say that recent criticisms by some workers in America, particularly that of Dr. A. M. Wenner and his associates, have led to a very careful re-assessment and repetition of the key experiments. It is very pleasant to be able now to assert without fear of contradiction that while von Frisch may have at times underestimated the importance of the search for odors when finding a food source, his main results have been overwhelmingly confirmed, particularly by the work of J. L. Gould who, by most ingenious experiments, has confirmed Karl's conclusions up to the hilt."[38] Thorpe's assessment of the debate was not uncommon.[39] Griffin, in his 1976 plea for a cognitive ethology, wrote of Wenner and Wells: "Their unwarranted skepticism has had the constructive effect of stimulating several new and improved experiments." He dismissed their critique by calling to "the ingenuity with which sufficiently determined skeptics can find some tortuous loophole providing for an indirect effect of odors." And ultimately, he concluded that "the weight of evidence is now overwhelmingly in favor of von Frisch's original interpretation."[40] When Wenner and two colleagues wrote an indignant letter to *Scientific American*, protesting that E. O. Wilson had made no mention of their challenge in his lengthy discussion of animal communication, Wilson replied, "I didn't mention the 'controversy' over the honeybee waggle dance because I no longer feel that the discussion of the subject even merits the word."[41] But within the same month that Wilson declared the debate dead, the biologist Demorest Davenport (and colleague in Wenner's department at the University of California, Santa Barbara) wrote a letter to *Science* in which he agreed with Stuart Altman that the debate was "one of the few nonsterile controversies in the study of animal behavior."[42]

What could have been taken as a compromise between the two sides was seen by most as von Frisch's victory. The casual reference to "Karl" in Thorpe's account of the debate also hints at the close personal ties and support von Frisch enjoyed on both sides of the Atlantic and Channel. Moreover, Wenner's insistence on odor and only odor hardly helped his cause. By 1973, the same year that von Frisch won a share of the Nobel Prize, Wenner had all but withdrawn from bee studies, citing a climate that had turned hostile to his contributions.[43]

But the focus on vindication over compromise also suggests that a great deal more was at stake than simply whether bees use odor or dance information to locate foods. Indeed, the debate touched

FIGURE 8.2. Karl von Frisch around the time he received the call from the Karolinska Institute in Sweden. He was eighty-six years old when he received the Nobel Prize. (Nachlaß Karl von Frisch, Bayerische Staatsbibliothek, Munich, ANA 540.)

on fundamental questions about the nature and capabilities of animals and how science ought to study them. In his *Science* piece, Gould recognized as much: "Some of the resistance to the idea that honey bees possess a symbolic language seems to have arisen from a conviction that 'lower' animals, and insects in particular, are too small and phylogenetically remote to be capable of 'complex behavior.'"[44] But complex behaviors were coming back into vogue during the 1960s and 1970s, and many saw von Frisch's work as an important arrow in the antibehaviorist's quiver.

The dispute over the bee dances was never just about whether the bees locate foods by means of a dance language or simply by reacting to odor cues. At stake were competing visions of animals and their behaviors and how science could best come to know them. Von Frisch neither fought nor won the battle by himself, as James Gould's decisive role in the debate suggests. His theory offered an interpretive flexibility that allowed it to be incorporated into a range of programs, such as the sociobiological perspective of E. O. Wilson or the more cognitive turn embraced by Donald Griffin. Although Wenner himself was no simple behaviorist, Griffin's and Mayr's reactions to the debate reveal that his resistance to von Frisch's theory was ascribed to precisely such tendencies. Indeed, von Frisch's case was picked up by a range of scientists reacting to the overly restrictive behaviorism that had dominated much of the century's American studies of behavior. And as it turned out, the bees could dance to a variety of tunes.

# 180/60

⚘

The Bavarian State Library in Munich stands in the heart of the city not far from where von Frisch's institute used to be. Stone steps lead past classical figures; Homer, Aristotle, Thucydides, and Hippocrates guard the entrance to this temple. Inside, the rumble of the street gives way to a cool, marbled silence. A grand stairway leads upstairs, where a special section is dedicated to the library's rare books and manuscripts. In an anteroom, visitors leave their coats and bags in metal lockers—only laptops, pencils, and notebooks are permitted. Inside the reading room, hushed voices discuss sources while gloved hands turn gorgeous pages of medieval manuscripts against the background tapping of fingers on laptops. Here one may also request the papers of Karl von Frisch.

The bulk of the von Frisch holdings in Munich consists of letters that span his earliest childhood cursive to the shaky scrawl of his twilight years. Hundreds of pages of typed carbon copies track his life, career, and family affairs over the decades. The collection also includes a set of journals.[1]

Historians cannot help but love journals—at their best, they can offer a unique view into a person's most private thoughts. But von Frisch, true to form, was not given to effusive revelations. Nonetheless, the books offer insights into aspects of his life that would have been difficult if not impossible to glean another way. His travel accounts, for example, contain candid reflections of the various people and places he encountered. Throughout most of his life, he also kept beautifully detailed notes on his research. He recorded wind and weather conditions, temperatures, and locations, as well as column after column of numbered bees. Together these items leave tantalizing trails of his scientific procedures and thoughts.

Among these journals lies another set of booklets that are far more intimate than the rest. For close to the last twenty years of his life, von Frisch kept records of his medical conditions. Doctors' visits, ailments and treatments, heart pains, and the surgeries that were to reverse his declining eyesight are captured in painstaking detail. Saddened and dulled by pages upon pages of chronic illness and suffering, I was unprepared for the final one of these notebooks.

The last journal covers just over two years and contains a continuous stream of dates and numbers recorded in shaky script. The final entry is dated June 5, 1982: 180/60. He died exactly one week later at the age of ninety-five. As I leaf through the notebook and take in these columns running the lengths of the pages, it dawns on me—these numbers are his blood pressure readings. And it is in these pages that I find encapsulated aspects of his personality that have struck me over the decade that I have researched and lived among the traces of his life and work. These numbers distill him into an essence far more profound than their physical referents— even as his life was about leave him, he dutifully recorded the phenomena with the scientist's unrelenting gaze.

In many ways, von Frisch's genius was his obsession. His tireless fascination with the natural world and his ability to pursue it to the exclusion of nearly everything else made him one of the most gifted scientists of the twentieth century. His experiments continue to be cited in scientific publications and draw open admiration from scientists for their rigor and elegance. They were ingeniously designed and meticulously executed. Sitting at observation stations, he and his assistants diligently recorded tens of thousands of bee numbers.

His work exhibited constant innovation and integration of different surroundings: for some trials, he and his assistants lured bees up nearby mountains; on another occasion, they rowed them to the middle of a lake; while for an elaborate experiment of the 1950s, a crate of bees was flown from Paris to New York by jet. His creativity in designing experiments was matched by a dedication that might put the bees' reputation to shame; each day he rose before sunrise to resume the work he had left the day before. On days that were too cold or rainy for the bees to leave the hive, he would write in his journal with palpable regret, "No experiments, bad weather," before he turned to indoor work.

This deep immersion in and commitment to his work go a long way toward explaining his fight to keep his position during the Nazi period. But to view it as merely the struggle of a man narrowly obsessed with his job would be to sell his life story short. For von Frisch, the changes under the regime and the threat of losing his work signaled the end of a world he had known since childhood through his parents, relatives, and family friends, all of whom were part of a lively intellectual community. By being declared partially Jewish, von Frisch came face-to-face with uncertainty about his livelihood and safety, especially as Nazi policy toward non-Aryans continued to sharpen over the course of the war. Still, we might ask what kind of a man was von Frisch in this time of crisis? What were his motives, fears, and hopes? And why should we care, when millions faced far greater persecution and suffering?

Von Frisch was no fan of the regime. He bridled at the control the party exercised over whom he could and could not hire, and he sought to keep Jewish employees in his lab as long as possible. The animal behaviorist (and future fellow laureate) Nikolaas Tinbergen worried as he witnessed von Frisch returning a passing student's "Heil Hitler" with a quiet "Grüss Gott." Before his own ancestry was called into question, he helped secure the release of a group of Polish scientists who had been interned at the Dachau concentration camp. But even together these actions do not amount to outright resistance to the regime. Why, then, should we care?

While it is tempting to slot the people of the time into roles as villains and heroes, much is to be gained by resisting the temptation. By looking beyond sharp binaries, such as collaborators and resistors, we get a more accurate and compelling portrayal of the time and lived experiences of most people under Nazism. By increasing our tolerance for this view without losing sight of the very real evil that existed at one end of the spectrum and the breathtaking courage and righteousness at the other, we can hope to gain a better understanding of how ordinary people could participate in a regime of such heinous intentions and consequences. Indeed, as some historians have argued, it was the coexistence of the "normal" alongside the murderous that allowed a regime initially thought by many to be a fly-by-night phenomenon to maintain its grip on power for twelve years.

If we look at German science in this period, a parallel story emerges. Much has been written about the troubling aspects of Nazi science, about scientists and intellectuals who willingly contributed ideological justifications for the regime's racial policies and actively participated in mass murder and human experimentation. It is sometimes hard to imagine that alongside such acts, there was also a good deal of science that was neither overtly ideological nor racially motivated and in many ways resembled the work that occurred in other nations involved in global conflict. And yet these "innocuous" German sciences also deserve our critical attention. For it was these areas—including plant and animal breeding, agriculture, and honeybee research—that received the most generous funding from the Nazi government. The regime counted on these sciences to provide food and munitions to wage and sustain its war effort. The dying bees, therefore, became critical to the Ministry of Food and Agriculture, and von Frisch's work must be seen as part of the regime's efforts to wage war. With this perspective in mind, the historical irony of von Frisch's position comes into stark relief— von Frisch was declared a Quarter Jew *and* performed his most important work with generous funding from the Nazi government.

While little is remembered today of the political circumstances from which the dance language emerged, von Frisch's designation as a Quarter Jew, in turn, helped him forge connections after the war. It allowed him to reestablish contacts that were important to both his funding and his weathering the sustained attack on his work in the 1960s and early 1970s. It was this steadfast support on both sides of the Atlantic that allowed him and his bees to assume such a central role in twentieth-century science.

But how did the bees become scientific superstars alongside

von Frisch? In some respects the insects had always occupied a privileged position in the history of science. From Aristotle's observation of their dances, to early microscopic investigations of their reproductive organs, to Darwin's musings on their comb-building instincts, the honeybees had been a favored scientific object through the ages. But something more profound happened through the work of Karl von Frisch. The honeybee dances emerged in the postwar period as the most studied form of nonhuman communication. For a brief period, the insects captured the rapt attentions of the public and researchers from various disciplines. Their dances were included in linguistics textbooks, those very works that were to train young scholars in the meaning of language, that ultimate human faculty that separates us from the beasts.

In the final chapter of the Cornell University bee researcher Thomas Seeley's *Honeybee Democracy*, entitled "Swarm Smarts," the author explains how the lessons of the hive can be applied to the human collective: "Fortunately, the house-hunting bees show us a brilliant solution to the puzzle of what gives rise to good group decision making. It is a solution that has been honed by natural selection for many millions of years . . . so it is certainly a time-tested method for achieving collective wisdom."[2]

While there is of course nothing new in looking to the hive for instruction, a striking difference emerges in this recent borrowing—reminiscent of von Frisch's own work, his intellectual grandson has demoted the queen. Seeley writes that the queen is no longer the "Royal Decider" but has instead become the "Royal Ovipositer." And her contribution to the hive is just that: "Each summer day, she monotonously lays the 1,500 or so eggs needed to maintain her colony's workforce. She is oblivious of her colony's ever-changing labor needs—for example, more comb builders here, fewer pollen foragers there—to which the colony's staff of worker bees steadily adapts itself." Thus, she is no longer afforded the decision-making role she enjoyed over the previous centuries. Instead, she has become the hive's reproductive machine and the "genetic heart of a swarm."[3] It is the collective that now teaches those prepared to heed the advice of the hive. And although its tens of thousands of members are driven by instinct, a kind of emergent intelligence prevails. The lessons Seeley offers are as sensible as they are intuitive. And yet the extrapolation from bee to human demands our pause. Why ought we to look to the hive for this wisdom?

There are no simple answers to the enduring lure of the hive.[4] Despite science's efforts to avoid anthropocentric projections onto animals, surely the questions we bring to them are very much our own.

The story of the bees resonates now more than ever as we face another kind of global crisis. In 2006, first reports began to trickle in of a condition that would soon capture the attention of the world. Entire honeybee colonies were found abandoned by their workers. Combs contained larvae that were still alive but would soon die of starvation, and undefended honey stores remained eerily untouched by bees from other hives or scavengers. Something was gravely wrong. Reports of dying bees kept mounting, and in 2007, 2008, and 2009, beekeepers reported over 30 percent losses in their colonies. In a recent "Progress Report on Colony Collapse Disorder," the US Department of Agriculture concluded that despite numerous theories, no single cause can be definitively linked to the disease.[5] Instead, the report calls it a "syndrome of stress" and adduces a range of possible co-contributors, including large-scale monoculture, inadequate nutrition of high fructose corn syrup, trucking the animals over long distances, and sustained pesticide use. The list of possible causes of this lethal stress cocktail reads as a poignant reminder of the limits to our ways of knowing and as a sobering indictment of our modern way of life. What will we do if the bees are no longer here to pollinate our food crops?

As the bees face an uncertain future, books and movies celebrate our renewed interest in them. Cities throughout the country are passing legislation to allow their urban dwellers to keep bees on rooftops and in gardens to satisfy our yearnings for communion with the hive. The famous entomologist E. O. Wilson—himself a professed ant rather than bee man—perhaps said it best when he called the bee "humanity's greatest friend among the insects."[6] And as is the case in great friendships, we are in this together—the bee and we, for better and worse.

# Acknowledgments

Many people have helped and contributed to this project over the years. I am grateful to Bob Richards, who introduced me to the history of science when I first landed in his heady graduate seminar at the University of Chicago. John Beatty, Sally Kohlstedt, and Jennifer Gunn are wonderful scholars and mentors, and I thank them for teaching me to think and write as a historian. Also from my Minnesota days, I thank Kai-Henrik Barth, Lisa Gannett, Mark Largent, and David Sepkoski. Conversations with Juliet Burba and Richard Werbowenko about life, work, museums, and horses have accompanied and delighted me over the years. At Princeton, I was fortunate to work with Michael Gordin, Mike Mahoney, and especially Angela Creager, whose warmth and encouragement have accompanied this project throughout. Angela was the one who insisted that the bees must fly! I am also grateful for the mentorship and support of Dan Todes and Chip Burkhardt; their scholarship and integrity have left a lasting impression, and Chip has generously read draft after draft of most of my work. I also thank Scott Curtis and Oliver Gaycken for their friendship and help with all things film. And to my lovely friends from Princeton, wafts of gratitude and nostalgia wash over me as I think of you—Daniela Bleichmar, Dan and Liz Bouk, Kerry Bystrom, Jamie Cohen-Cole, Liz Foster, Katie Holt, Molly Loberg, Sinead MacNamara, Jane Murphy, Carla Nappi, Cat Nisbett Becker, Joe November, Katrina Olds, Jeff Schwegman, Suman Seth, Mitra Sharafi, Alistair Sponsel, Laura Stark, Karen Velez, Jenny Weber, and Matt Wisnioski. I am especially grateful to my dear friends Daniela, Jane, Liz, Jenny, Molly, and Suman for their love and support throughout all of it. Helen Tilley has been a wonderful and generous comrade in arms, and I feel deeply fortunate that our paths have crossed again. Her friendship, generosity, and wisdom are deeply appreciated.

I thank Klaus Taschwer, Oliver Hochadel, Benedict Föger, Veronika Hofer, and Mitch Ash in Vienna. Klaus's unstinting support and generosity over the years have put me deep in his debt; our chance meeting in a musty basement in Altenberg was fortunate indeed.

From my Berlin days, I thank Nancy Anderson, Robyn Braun, Matthias Brun, Mirjam Brusius, Suparna Choudhury, Philip Kitcher, Steffi Klamm, Fabian Krämer, Erika Milam, Gregg Mitman, Christine von Oertzen, Jeff Schwegman, Thomas Sturm, Fernando Vidal, Annette Vogt, and especially Veronika Lipphardt. Janka Deicke, David Ludwig, and Jack Quigley were able assistants. I thank Lorraine Daston for inviting me into the queendom of the mind and for making it a place worthy of *Heimweh*. And Conny Schoop for her friendship and unstinting support during Berlin and after.

At Northwestern, I thank Ken Alder, Tasha Dennison, and Steve Epstein in Science in Human Culture. I am especially glad to have Alice Weinreb and Edin Hajdarpasic in my life. I am extremely fortunate to have found my way into the Northwestern advising office; I could not have asked for more wonderful colleagues! I thank friends and supporters in Chicago: Chris Bell, Wendy Boxer, Janina Cabizza, Deborah Cohen, Mariana Craciun and Camilo Leslie, Jack Donohew, Sheila Donohue, Leo Lane, Hilarie Lieb, Micki O'Neill, Fay Rosner, Priya Shankar, Dan Stolz, Barbara Tilley, Liz Trubey, Jennifer Wells, and Shirley and Grant Zimmerman.

Audra Wolfe is the smart outside reader who helped with this manuscript at critical stages, as did Michael Maltenfort. Tom Seeley lent the project his scientist's eye, and I thank them all for catching and generously correcting many hiccups. The remaining ones are mine.

I gratefully acknowledge the support of the National Science Foundation, Princeton's Center for Human Values, and the Max Planck Society. Thanks to the von Frisch family, especially Julian von Frisch, for granting me permission to publish from the beautiful collection of Karl von Frisch papers at the Bayerische Staatsbibliothek in Munich. Jeann Healey and Lore Becker shared with me their reflections of what it was like to work with Karl von Frisch. I am grateful to the many archive and interlibrary loan wizards who contributed to this work—they waved their wands on both sides of the Atlantic and were especially kind to me in Princeton, Berlin, Vienna, Evanston, and Munich, where I spent many months. At the University of Chicago Press, Kelly Finefrock-Creed did a beautiful job copyediting the manuscript, and I am deeply grateful to Karen Darling, who believed in this project from the very beginning and made its current form possible. I am thrilled she gave it this home.

My dear friends Amy Doolan Roy and Emily Bremner Forbes—you are the best, and I love and adore you. I also thank Anna Stelow, Emily Ashton, Elisa Vargas, Elizabeth Esse, Eva Nielsen, Gina Fridland, Heather Dunbar, and Nina Baumgartner for their love and support over the years and across continents. Ruth Widmer, Hans and Marjolein Koppen, Lizzie and Stirling Stackhouse, Helen Tozer, and Christa, Rolf, Désirée, and Roman Akermann have all been wonderful friends to my family.

And to my family, I owe the greatest debt. I am deeply grateful to my sister and brother-in-law, Beatrice and Jimmy Hallac, for their steadfast love and support. My sister, especially, has endured much thick description from the book-writing trenches and its peripheries. And to my nieces Anna, Maya, and Sophia—so funny, smart, and wonderful; we are so lucky to have you in our lives! My parents Rudi and Aurora Munz made it all possible. My mom read the manuscript top to bottom when it was still far from finding its way between covers. My dad did not live to see its publication. But I know he would have been very proud, and that fills me with joy and sadness. It is to them that I dedicate this book. And to Tim O'Leary, who entered the thicket and held my hand as I slogged across the finish line. Thank you for your love, for your humor, and for believing that things are possible. They are.

# Notes

## Introduction

1. Karl von Frisch to Otto Koehler, January 12, 1946, Nachlaß Karl von Frisch, Bayerische Staatsbibliothek, Munich, ANA 540 B.II. Throughout the book, unless otherwise noted, all translations from German into English are by the author.

2. For example, the linguist Thomas Sebeok organized an international conference on animal communication that took place in June of 1965 in Austria. It was sponsored by the Wenner-Gren Foundation for Anthropological Research and resulted in two influential volumes: Sebeok, *Animal Communication*; and Sebeok and Ramsay, *Approaches to Animal Communication*.

3. Bee communication was deemed the most widely studied nonhuman form of communication by Wenner in "Honey Bees" (1968) and Altman in "Without a Word." For references to the bee language's high level of complexity, see Haldane and Spurway, "A Statistical Analysis." When the term "language" is used in this book in reference to the bee dances, it is with the understanding that the bee dances do not satisfy what linguists would refer to when they use "language" as a technical term. I use it throughout this book in reference to the historical term "honeybee dance language."

4. Little scholarly work has been produced on Karl von Frisch despite his importance to science and animal behavior studies. A good starting point is his autobiography, *A Biologist Remembers*. In this book, I use the German original, *Erinnerungen*. See also Kreutzer, *Karl von Frisch*; and Burkhardt, "Karl von Frisch."

5. For material related to von Frisch's presentations and talks, see call numbers ANA 540 A.V, ANA 540 A.VI, and ANA 540 B.VI.8 within the Nachlaß Karl von Frisch collection at the Bayerische Staatsbibliothek, Munich.

6. For a historical discussion of moral projections onto nature, see Daston and Vidal, *The Moral Authority of Nature*. For bees in particular, see Merrick, "Royal Bees"; and Allen, "Burning the Fable of the Bees."

7. The literature on this topic is vast and often specific to the institution, location, and industry in question. For a general overview, see Bracher, "Stages of Totalitarian 'Integration' (*Gleichschaltung*)"; Evans, *The Third Reich in Power*; and Burleigh, *The Third Reich*.

8. Elon, *The Pity of It All*, chs. 7–8.

9. See Meyer, *"Jüdische Mischlinge"*; and Benz, *Die Juden in Deutschland*.

10. Von Frisch, *Erinnerungen*, 115.

11. See Corni and Gies, *Brot, Butter, Kanonen*.

12. See Götz and Heim, *Vordenker der Vernichtung*; Heim, "Research for Autarky"; and Saraiva and Wise, "Autarky/Autarchy."

13. A. D. Hasler of the Rockefeller Foundation judged that von Frisch's politics were sound. Arthur Hasler to George H. Parker and Dwight Minnich, August 25, 1945, and Arthur Hasler to Raymond B. Fosdick, August 26, 1945, Rockefeller Foundation Archive, Sleepy Hollow, New York, RG 1.1, Series 717D, Box 14, Folder 136 (University of Munich 1948–1952). See also H. Bentley Glass's 1951 assessment that he sees no reasons to doubt von Frisch's "commitment to democratic ideals." H. Bentley Glass, confidential internal memo, January 10, 1951, National Archives and Records Administration, RG 466 (HICOG), E-139 (Subject Files of the Scientific Research Division, Military Security Board), Box 15, Folder "Glass, H. Bentley–Report." I thank Audra Wolfe for bringing this to my attention.

14. Investigations into colony collapse disorder or syndrome are ongoing. For a selection, see Jacobsen, *Fruitless Fall*; Renée Johnson, "Honey Bee Colony Collapse Disorder," in Congressional Research Service Report for Congress, January 7, 2010, accessed August 23, 2015, http://www.fas.org/sgp/crs/misc/RL33938.pdf; and United States Department of Agriculture, *Colony Collapse Disorder Progress Report*, June 2010, accessed March 16, 2015, www.ars.usda.gov/is/br/ccd/ccdprogressreport2010.pdf.

### Bee Vignette I

1. Darwin, *Origin of Species*, 259; and Maclaurin, "On the Bases of the Cells," 571.

2. Reid, *The Works of Thomas Reid*, 546–47.

3. Darwin, *Origin of Species*, 259–66.

4. Romanes, *Animal Intelligence*, 176 (quotes), 177, 186.

5. Ibid., 169.

6. John Lubbock quoted in ibid., 156.

7. Ludwig Büchner quoted in ibid., 159.

8. John Lubbock quoted in ibid.

9. Josiah Emery quoted in ibid., 157–58.

### Chapter One

1. Von Frisch, *Fünf Häuser*, 6–9.

2. Marie Exner von Frisch quoted in Coen, *Vienna*, 72.

3. Von Frisch, *Fünf Häuser*, 8.

4. Ibid., 9.

5. Their time in Zurich is discussed in Coen, *Vienna*, 82–89.

6. The correspondence is published in Keller, von Frisch-Exner, Exner, and Smidt, *Aus Gottfried Kellers glücklicher Zeit*.
7. Von Frisch, *Fünf Häuser*, 10–11.
8. Ibid., 11.
9. Hatzinger, "Anton Ritter von Frisch," 719. Anton von Frisch was one of the thirty-eight founders of the German Society of Urology and its first president in 1907.
10. For a discussion of Brunnwinkl, see von Frisch, *Fünf Häuser*; and Coen, *Vienna*.
11. Von Frisch, *A Biologist*, 17.
12. Ibid., 25.
13. Ibid., 20.
14. Ibid., 24.
15. He recorded the various species in what he later deemed "a rather scrappy diary." The species counts were listed as follows: 9 mammals, 16 birds, 26 cold-blooded terrestrial vertebrates, 27 fish, 45 invertebrates. Ibid., 19.
16. Ibid., 19, 21–22 (quote). Decades later, Tschocki would also make an important appearance in a letter to the cognitive ethologist Donald Griffin. Von Frisch recounted the animal's intelligence in contradistinction to that of the honeybees. Karl von Frisch to Donald R. Griffin, November 13, 1976, Nachlaß Karl von Frisch, Bayerische Staatsbibliothek, Munich, ANA 540 B.II.
17. Ackerl, *Vienna Modernism*, 5; and Schorske, *Fin-de-Siècle Vienna*.
18. Coen, *Vienna*, 13; and Coen, "A Lens of Many Facets."
19. Von Frisch, *Erinnerungen*, 26.
20. Von Frisch, "Studien über Pigmentverschiebung." Von Frisch deemed Exner's work on the topic of such enduring significance that he initiated a republication and translation of the book as *The Physiology of the Compound Eyes of Insects and Crustaceans* in 1989, for which he wrote the introduction.
21. Von Frisch, "Studien über Pigmentverschiebung," 662.
22. Ibid., 701.
23. Von Frisch, *A Biologist*, 33 (quotes); and Von Frisch, *Erinnerungen*, 26–27.
24. Nyhart, *Biology Takes Form*, 157–59; and Von Frisch, *A Biologist*, 35–36. The book is Hertwig and Hertwig, *Die Actinien*.
25. Von Frisch, "Erinnerungen an Otto Koehler," 465.
26. Ibid., 466, 467.
27. Von Frisch, *Erinnerungen*, 30.
28. On the BVA, see Przibram, "Die Biologische Versuchsanstalt in Wien . . . (1902–1907)"; Przibram, "Die Biologische Versuchsanstalt in Wien . . . (1908–1912)"; Reiter, "Zerstört und Vergessen"; Coen, "Living Precisely in Fin-de-Siècle Vienna"; Hofer, "Rudolf Goldscheid, Paul Kammerer und die Biologen des Prater-Vivariums i"; Koestler, *The Case of the Midwife Toad*; and Brauckmann and Müller, *A Laboratory in the Prater*.
29. Taschwer, *Hochberg*, ch. 1; Taschwer, "From the Aquarium."
30. Przibram, "Die biologische Versuchsanstalt in Wien . . . (1902–1907)," 234.
31. Ibid., 246, 249, 258, 261.
32. Von Frisch, *Erinnerungen*, 33.

33. Ibid.
34. Von Frisch, "Beiträge zur Physiologie der Pigmentzellen," 323, 381.
35. Von Frisch, *A Biologist*, 42.

## Chapter Two

1. Karl von Frisch to Anton von Frisch, February 21, 1912, Nachlaß Karl von Frisch, Bayerische Staatsbibliothek, Munich, ANA 540 B.II.
2. Von Hess, "Untersuchungen über den Lichtsinn bei Fischen," 2, 3, 35.
3. Von Frisch, "Über den Farbensinn der Fische," 220.
4. Ibid., 222.
5. Ibid., 223–25.
6. Ibid., 224–25.
7. The article was published in German as "Untersuchungen zur Frage nach dem Vorkommen von Farbensinn bei Fischen."
8. Von Hess, "Untersuchungen zur Frage," 630, 634n (emphasis von Hess).
9. Ibid., 634n.
10. Ibid., 635–36, 645.
11. See Turner, *In the Eye's Mind*.
12. Exner and Exner, "Die physikalischen Grundlagen der Blütenfärbungen."
13. The article was published in German as "Sind die Fische Farbenblind?"
14. Von Frisch, "Sind die Fische Farbenblind?," 109–10.
15. Ibid., 112, 111, respectively.
16. Ibid., 126.
17. The article was published in German as "Weitere Untersuchungen über den Farbensinn der Fische."
18. Von Frisch adapted the conditioning technique from von Hess's "Experimentelle Untersuchungen," and "Über farbige Anpassung bei Fischen" describes von Frisch's first use of it.
19. Von Frisch, "Weitere Untersuchungen," 48.
20. Ibid., 48–49.
21. Von Hess, "Neue Untersuchungen," 407.
22. Von Frisch, *Erinnerungen*, 40.
23. Von Frisch, "Weitere Untersuchungen," 44n, 53.
24. Karl von Frisch to Anton von Frisch, September 18, 1913, Nachlaß Karl von Frisch, Bayerische Staatsbibliothek, Munich, ANA 540 C.III "Universitätsangelegenheiten," Mappe "Kontroverse K. v. Frisch-Carl Hess."
25. Von Frisch, "Weitere Untersuchungen," 67.
26. Von Hess, "Der Gesichtssinn," 660–69.
27. Ibid., 660–70.
28. Published in German as *Das entdeckte Geheimnis der Natur im Bau und in der Befruchtung der Blumen*.
29. On the moral economy and authority of nature, see Daston and Vidal, *The Moral Authority of Nature*.
30. Von Frisch, "Über den Frabensinn der Bienen," 16.

31. Ibid., 16.
32. Ibid.
33. Ibid., 17.
34. Ibid.
35. Ibid.
36. Ibid., 16.
37. On virtual witnessing technologies and visual proof, see Shapin, "Pump and Circumstance"; Shapin and Schaffer, *Leviathan and the Air-Pump*; and Daston and Galison, "The Image of Objectivity." Kirby has usefully extended Shapin and Schaffer's concept of the virtual witness to film. Kirby, "Science Consultants, Fictional Films, and Scientific Practice," 235.
38. Doflein, "Der angebliche Farbensinn der Insekten," 709.
39. See Lavédrine and Gandolfo, *The Lumière Autochrome*.
40. Doflein, "Der angebliche Farbensinn der Insekten," 710. Stellwaag, "Über die Beziehung des Lebens zum Licht," 1642.
41. Von Hess, "Beiträge zur Frage," 364–65.
42. Von Frisch, *Fünf Häuser*, 80 (quote); von Frisch, *A Biologist*, 61.
43. Von Frisch, *A Biologist*, 63.
44. Von Frisch, *Erinnerungen*, 55.
45. The manual was titled *Sechs Vorträge über Bakteriologie für Krankenschwestern*.
46. Von Frisch, *Fünf Häuser*, 83.
47. Ibid.
48. Von Frisch, "Über den Geruchsinn der Bienen," 6–8.
49. Here I follow von Frisch, "Über den Geruchsinn der Bienen," esp. 13–15, 19–23, 24, 25, 29, 46 (quote), 157 (quote).
50. The cardboard boxes were no longer available during the war. The ceramic boxes also had a 1.5 cm hole in the front and a lid that could be removed and replaced, and were slightly larger than the cardboard boxes had been (11 × 11 × 11 cm).
51. Von Frisch, *Erinnerungen*, 61.
52. Ibid., 60, 61 (quote).
53. Von Frisch introduced this work in a three-part series titled "Über die 'Sprache' der Bienen." The quote is from "Über die 'Sprache' der Bienen, I," 567.
54. Von Frisch, "Über die 'Sprache' der Bienen, II," 510.
55. Von Frisch, "Über die 'Sprache,' der Bienen, III," 781.
56. Von Frisch, "Über die 'Sprache' der Bienen, II," 509.
57. Von Frisch, *Über die "Sprache" der Bienen*, 73.
58. Von Frisch, "Über die 'Sprache' der Bienen, I," 568–69.
59. Von Frisch, *Über die "Sprache" der Bienen*, 81. And indeed, today bee researchers do not distinguish between these dances. See Gardner, Seeley, and Calderone, "Do Honeybees."
60. Von Frisch, *Über die "Sprache" der Bienen*, 10.
61. For a discussion of cultural investments in nineteenth-century works on ants and bees, see Clark, "'The Complete Biography of Every Animal.'" On

insect studies and gender, see Drouin, "L'Image des Sociétés d'Insectes"; and Prete, "Can Females Rule the Hive?" For a historical work, see Sleigh, *Six Legs Better*.

62. Von Frisch, *Aus dem Leben der Bienen* (1927), 2 (emphasis von Frisch.).

63. Methods for marking bees had been devised by other scientists. See, for example, John Lubbock's 1872–1879 diary in Clark, "'The Complete Biography of Every Animal,'" 263n52. But von Frisch's system allowed him to keep track of much larger numbers of bees. Moreover, his use of powder paint dissolved in a shellac-alcohol solution made for much longer lasting markings than the traditionally used oil paints. Von Frisch, *Über die "Sprache" der Bienen*, 21–24.

64. Von Frisch, "Über die 'Sprache' der Bienen, I," 567.

65. Von Frisch, *Erinnerungen*, 65 (quote), 67 (quote), 120.

66. Karl von Frisch to Marie Exner von Frisch, May 22, 1922, Nachlaß Karl von Frisch, Bayerische Staatsbibliothek, Munich, ANA 540 B.II.

67. Karl von Frisch to Marie Exner von Frisch, January 31, 1925, Nachlaß Karl von Frisch, Bayerische Staatsbibliothek, Munich, ANA 540 B.II.

## Bee Vignette II

1. For a select list of von Frisch's films of the 1920s and 1930s, see the film section of the bibliography.

2. For examples, see von Frisch, *Der Farbensinn und Formensinn der Biene*; von Frisch, "Über den Geruchssinn der Bienen"; and von Frisch, "Ein Zwergwels."

3. Von Frisch, *Geschmacksinn bei Fischen* [Sense of taste in fish]. A version of this section was previously published as Munz, "Die Ethologie des wissenschaflichen Cineasten."

4. Karl von Frisch, February 16, 1964, Nachlaß Karl von Frisch, Bayerische Staatsbibliothek, Munich, ANA 540 B.I, Mappe "Institut für den Wissenschaftlichen Film, Göttingen."

5. Karl von Frisch, July 21, 1958, Nachlaß Karl von Frisch, Bayerische Staatsbibliothek, Munich, ANA 540 B.I, Mappe "Österreichischer Rundfunk, Wien."

6. Von Frisch, *Erinnerungen*, 82.

7. See MacLuhan, *The Medium Is the Message*.

8. Danto, "Giotto and the Stench of Lazarus." I am grateful to Edward Eigen for bringing this article to my attention.

9. He did this to prevent the animal from being cued visually. Von Frisch, "Über den Gehörsinn der Fische," 1.

10. See Braun and Marey, *Picturing Time*; Dagognet, *Etienne-Jules Marey*; Prodger, *Time Stands Still*; and Solnit, *River of Shadows*.

11. Burt, *Animals in Film*, ch. 1.

12. Ibid.

## Chapter Three

1. Heinroth, *Mit Faltern begann es*, 74.
2. Von Frisch and Kollmann, *Der Neubau*, 4; "Lehrstuhlbesetzungen, Liste der bei von Frisch entstandenen Dissertationen," Nachlaß Karl von Frisch, Bayerische Staatsbibliothek, Munich, ANA 540 C.III "Universitätsangelegenheiten," item 77; Von Frisch, *Erinnerungen*, 96.
3. See Gemelli and MacLeod, *American Foundations in Europe*; and Giuliana Gemelli, Picard, and Schneider, *Managing Medical Research in Europe*.
4. Von Frisch and Kollmann, *Der Neubau*, 4.
5. For an account of the negotiations, see Lauder Jones, "Report re Institute of Zoology and Institute of Physical Chemistry of the University of Munich," March 10, 1930, Rockefeller Foundation Archive, Sleepy Hollow, New York, RG 1.1, Series 717D, Box 14, Folder 134.
6. Von Frisch, *Erinnerungen*, 98.
7. Ross G. Harrison to the Rockefeller Foundation, May 18, 1929, Rockefeller Foundation Archive, Sleepy Hollow, New York, RG 1.1, Series 717D, Box 14, Folder 133
8. Von Frisch, *Erinnerungen*, 99.
9. Ibid., 100.
10. Sinclair, *The Jungle*, 38, 39.
11. Von Frisch, *Erinnerungen*, 100.
12. Ibid., 100–101.
13. Karl von Frisch to the Foreign Office (Auswärtiges Amt) in Berlin, via the Bavarian State Ministry for Education and Cultural Affairs, May 23, 1930, Bayerisches Hauptstaatsarchiv, Munich, MK 54482 (Personalakt Karl von Frisch) (emphasis von Frisch).
14. Karl von Frisch to Margarete von Frisch, March 28, 1930, Nachlaß Karl von Frisch, Bayerische Staatsbibliothek, Munich, ANA 540 B.II.
15. Karl von Frisch to the Foreign Office (Auswärtiges Amt) in Berlin, via the Bavarian State Ministry for Education and Cultural Affairs, May 23, 1930, Bayerisches Hauptstaatsarchiv, Munich, MK 54482 (Personalakt Karl von Frisch).
16. Meeting minutes, June 17, 1949, Rockefeller Foundation Archive, Sleepy Hollow, New York, RG 1.2, Series 705D, Box 6, Folder 56.
17. Von Frisch, *Erinnerungen*, 105; and von Frisch and Kollmann, *Der Neubau*, 1–32.
18. Böhm, *Von der Selbstverwaltung*, 58–84. For a parallel discussion of how "coordination," or *Gleichschaltung*, occurred at the University of Heidelberg, see Remy, *The Heidelberg Myth*, esp. chs. 1–3.
19. Böhm, *Von der Selbstverwaltung*, 27–57; and Litten, *Der Rücktritt Richard Willstätters*.
20. Böhm, *Von der Selbstverwaltung*, 130, 426–94.
21. The full title of the law was the Restoration of the Professional Civil Service Law (Gesetz zur Wiederherstellung des Berufsbeamtentums). "The legislation commenced with the Law for the Restoration of the Professional Civ-

il Service of 7 April 1933, followed by measures against Jewish physicians, teachers, and students, on the 22nd and 25th of the same month. The former sanctioned the dismissal of both the undesirable and 'non-Aryans' from the public service; the latter attempted to either remove Jews from or to restrict access to the professions, while encouraging 'Aryans' to dispense with the services of Jews." Burleigh and Wippermann, *The Racial State*, 44.

22. "Fragebogen zur Durchführung des Gesetzes zur Wiederherstellung des Berufsbeamtentums vom 7. April 1933," June 14, 1933, Personalakt Karl von Frisch, Ludwig Maximilians University Archive, Munich, Folder E-II-1376 (Akten des Senats).

23. Böhm, *Von der Selbstverwaltung*, 130. By comparison, Klaus Hentschel estimates that 15 percent of all university teachers were suspended in 1933/34 as a result of the new laws. Hentschel, introduction to *Physics and National Socialism*, lv.

24. Joann Healey, interview by the author, London, May 28, 2010. In a postwar letter to Richard Goldschmidt, von Frisch also lists members of the faculty who were considered "unsafe" because of their Nazi sympathies. Karl von Frisch to Richard Goldschmidt, April 5, 1946, Nachlaß Karl von Frisch, Bayerische Staatsbibliothek, Munich, ANA 540 B.IV.

25. Large, *Where Ghosts Walked*, 238. The denunciation fervor among Germans under the Nazis is analyzed by Gellately, *The Gestapo and German Society*.

26. See Sax, *Animals in the Third Reich*; Klueting, "Die gesetzliche Regelung"; and Smith, "'Cruelty of the Worst Kind.'" For the legislation enacted to protect animals, see Giese and Zschiesche, *Die deutsche Tierschutzgesetzgebung*. An English translation of the law may be found in Sax, *Animals in the Third Reich*, 175–79.

27. Von Frisch, *Erinnerungen*, 114.

28. "Der neutrale Gelehrte," *Deutsche Studentenzeitung, Kampfblatt der NS-Studenten* (Munich), December 6, 1934, 6.

29. Rektorat, Biologisches Institut Universität München, Anlage I, April 6, 1936, Personalakt Karl von Frisch, Ludwig Maximilians University Archive, Munich, Folder E-II-1376 (Akten des Senats).

30. Nagel, "'Er ist der Schrecken überhaupt der Hochschule'"; Böhm, *Von der Selbstverwaltung*, 553.

31. Dr. Führer of the Dozentenbund to the Rector of the University of Munich, April 6, 1936, Personalakt Karl von Frisch, Ludwig Maximilians University Archive, Munich, Folder E-II-1376 (Akten des Senats).

32. Bundesarchiv, Berlin Lichterfelde, Reichssippenamt, RSA, R 1509-2, File "Karl von Frisch."

33. On the emergence of European popular science writing as a genre, see Daum, *Wissenschaftspopularisierung im 19. Jahrhundert*; and Azzouni, "Wissenschaftspopularisierung um 1900."

34. Bundesarchiv, Berlin Lichterfelde, NS 15 36, 81a.

35. *The Dancing Bees* would go through ten editions in total, the last of which was published posthumously in 1993.

36. Von Frisch, *Du und das Leben* (Deutscher).

37. Bundesarchiv, Berlin Lichterfelde, NS 15 36; 81a, February 17, 1938.

38. "Deutsches Beamtengesetz, January 26, 1937," accessed March 16, 2015,

http://www.verfassungen.de/de/de33-45/beamte37.htm. The quoted material is from para. II, 1. "Allgemein," §3 (1), and para. II, 2. "Treueid," §4. (1).

39. For more detail, see Burleigh and Wippermann, *Racial State*, 44.
40. Von Frisch, "Psychologie der Bienen."
41. Twenty-one of the thirty pieces cited were published in the journal.
42. One of von Frisch's students, Gustav Rösch, designed and carried out a painstaking longitudinal study of the bees' habits for which he carefully numbered freshly hatched bees and followed their activities through the course of their lives. He published his work in three separate papers: "Untersuchungen über die Arbeitsteilung im Bienenstaat I" (1925); "Über die Bautätigkeit im Bienenvolk und das Alter der Baubienen" (1927); and "Untersuchungen über die Arbeitsteilung im Bienenstaat II" (1930).
43. Von Frisch, "Psychologie der Bienen," 9–10.
44. Rösch, "Über die Bautätigkeit," 12–14; and von Frisch, "Psychologie der Bienen," 12 (quotes).
45. Kressley-Mba and Jaeger, "Rediscovering a Missing Link."
46. Von Frisch, "Psychologie der Bienen," 13, 14.
47. Ibid., 14.
48. Ibid., 16, 18.
49. Ibid., 12, 20.
50. Karl von Frisch to Curt Stern, January 12, 1933, Curt Stern Papers, American Philosophical Society, Philadelphia, PA, Box 9.
51. For an excerpt of Stern's letter, see Deichmann, *Biologists under Hitler*, 21.
52. Deichmann, *Biologists under Hitler*, 20–21; and Von Frisch, *Erinnerungen*, 113.
53. Vogt, *Wissenschaftlerinnen in Kaiser-Wilhelm-Instituten*, 62.
54. Hertz never recovered from having been forced to leave her homeland. She never applied for British citizenship and was reported to have died alone, partially blind, and relatively impoverished in 1975. Jaeger, "Vom erklärbaren, doch ungeklärten Abbruch einer Karriere."
55. I base this account on the excellent work by Seyfarth and Perzchala, "Sonderaktion Krakau 1939." The von Frisch quotes are my translations of quoted material found on page 222. I translate and quote additional material from pages 222 and 223.
56. Jan Bayer to Rektor der Universität, December 23, 1940, Personalakt Karl von Frisch, Ludwig Maximilians University Archive, Munich, Folder E-II-1376 (Akten des Senats).

### Chapter Four

1. Karl von Frisch to Otto Koehler, January 11, 1941, Nachlaß Karl von Frisch, Bayerische Staatsbibliothek, Munich, ANA 540 C.II.13 "Abstammungsbescheid" (hereafter designated Nachlaß Karl von Frisch).
2. Deichmann, *Biologists under Hitler*, esp. 90, 98–99, 117, 119–20, 161, 163.
3. Hans Spemann quoted in von Frisch, *Erinnerungen*, 65.
4. Hans Spemann to Karl von Frisch, January 12, 1941, Nachlaß Karl von Frisch, ANA 540 C.II.13 "Abstammungsbescheid."

5. Hans Spemann to Bernhard Rust, January 21, 1941, Nachlaß Karl von Frisch, ANA 540 C.II.13 "Abstammungsbescheid."
6. Otto Koehler to Karl von Frisch, April 1, 1942, Nachlaß Karl von Frisch, ANA 540 C.II.13 "Abstammungsbescheid."
7. Karl von Frisch, "Deutsche Forschung im Krieg," *Das Reich*, February 2, 1941, 19.
8. Von Frisch, "Psychologie der Bienen."
9. Weinreb, *Modern Hungers*.
10. See Hopke, *Medizin und Verbrechen*.
11. See Falk, *Genetic Analysis*; and Müller-Wille and Rehinberger, *Heredity Produced*.
12. For discussions of eugenics, see Kevles, *In the Name of Eugenics*; Weindling, *Health, Race, and German Politics*; and Paul, *Controlling Human Heredity*.
13. Von Frisch, *Du und das Leben* (Berlin: Deutscher Verlag, Dr.-Goebbels-Spende für die deutsche Wehrmacht, 1936); von Frisch, *Du und das Leben: Eine Moderne Biologie für Jedermann* (Berlin: Ullstein Verlag, 1936).
14. Von Frisch, *Du und das Leben* (Deutscher), 358.
15. Von Frisch, *Du und das Leben* (Ullstein), 345, 346.
16. See Bashford and Levine, *Oxford Handbook of the History of Eugenics*; Adams, *The Wellborn Science*; and Lipphardt, *Biologie der Juden*.
17. Von Frisch, *Du und das Leben* (Deutscher), 362.
18. Otto Koehler to Karl von Frisch, February 4, 1941, Nachlaß Karl von Frisch, ANA 540 C.II.13 "Abstammungsbescheid."
19. Jan Bayer to Rektor der Universität, December 23, 1940, Personalakt Karl von Frisch, Ludwig Maximilians University Archive, Munich, Folder E-II-1376 (Akten des Senats); Karl von Frisch to Otto Koehler, January 11, 1941, Nachlaß Karl von Frisch, ANA 540 C.II.13 "Abstammungsbescheid"; and Otto Koehler to Karl von Frisch, January 20, 1941, Nachlaß Karl von Frisch, ANA 540 C.II.13 "Abstammungsbescheid" (quote).
20. Karl von Frisch writes about his brother's unwillingness to talk about his leaving the party in Karl von Frisch to Fritz von Wettstein, January 20, 1941, Nachlaß Karl von Frisch, ANA 540 C.II.13 "Abstammungsbescheid."
21. The following details are related by Hans von Frisch to Karl von Frisch, January 26, 1941, Nachlaß Karl von Frisch, ANA 540 C.II.13 "Abstammungsbescheid."
22. Otto Koehler to Karl von Frisch, January 27, 1941, Nachlaß Karl von Frisch, ANA 540 C.II.13 "Abstammungsbescheid."
23. The following events are recounted in Alfred Kühn to Karl von Frisch, February 6, 1941, Nachlaß Karl von Frisch, ANA 540 C.II.13 "Abstammungsbescheid." The quoted material in the next two paragraphs is from this letter.
24. Alfred Kühn to Karl von Frisch, February 6, 1941, Nachlaß Karl von Frisch, ANA 540, C.II.13 "Abstammungsbescheid."
25. Decker to Rektor der Universität, February 10, 1941, Nachlaß Karl von Frisch, ANA 540 C.II.13 "Abstammungsbescheid."
26. There is no study, so far as I know, on *Gleichstellungen* (equalization) for the civil service. Similar requests were made in cases of ambiguous ancestry, where individuals requested to be equalized to Aryans. Such requests had

to be granted by Hitler personally and became increasingly rare as the war continued. Meyer, *"Jüdische Mischlinge."*

27. Von Frisch's notes about these meetings are written on a piece of paper with the title "Ergebniße d. Berlin Reise, 18.–20.II.41" (Nachlaß Karl von Frisch, ANA 540 C.II.13 "Abstammungsbescheid"). The subsequent discussion is based on this source.

28. Beyerchen. *Scientists under Hitler.*

29. Deichmann, *Biologists under Hitler*, 90.

30. That he met with Dr. Führer may be inferred from Karl von Frisch to Ferdinand Springer, February 21, 1941, Nachlaß Karl von Frisch, ANA 540 C.II.13 "Abstammungsbescheid."

31. In his notes of the day's meetings, he added an exclamation mark after the statement about the acknowledged value of his work.

32. Karl von Frisch to Ferdinand Springer, February 21, 1941, Nachlaß Karl von Frisch, ANA 540 C.II.13 "Abstammungsbescheid."

33. The statement is dated a day after the deadline, which is surprising, and the reason why is unclear. Karl von Frisch to Rektor der Universität München, March 11, 1941, Nachlaß Karl von Frisch, ANA 540 C.II.13 "Abstammungsbescheid."

34. Karl von Frisch to Hans Spemann, June 15, 1941, Nachlaß Karl von Frisch, ANA 540 C.II.13 "Abstammungsbescheid"; Hans Spemann to Karl von Frisch, June 17, 1941, Nachlaß Karl von Frisch, ANA 540 C.II.13 "Abstammungsbescheid."

35. Otto Koehler to Karl von Frisch, March 3, 1941, Nachlaß Karl von Frisch, ANA 540 C.II.13 "Abstammungsbescheid."

36. Von Frisch, *Fünf Häuser*, 103.

37. Corni and Gies, *Brot*, ch. 4.

38. For discussions and results of this work, see von Frisch, *Erinnerungen*, 116; von Frisch, *Fünf Häuser*, 103–4; von Frisch, "Bericht." 16: 12 (1943).

39. Von Frisch, *Erinnerungen*, 116.

40. On Franz Wirz's role in the food policies of the Third Reich, see Melzer, *Vollwerternährung*, esp. p. 183–98.

41. Von Frisch, *Erinnerungen*, 116.

42. Von Frisch, "Bericht."

43. Von Frisch, "Die 'Sprache' der Bienen," 403–4.

44. Alfred Kühn to Karl von Frisch, August 25, 1941, Nachlaß Karl von Frisch, ANA 540 C.II.13 "Abstammungsbescheid."

45. Ernst Bergdolt to Max Demmel, Oberregierungsrat of the Reich Ministry of Science, Education, and Public Instruction, May 21, 1941, Personalakt Karl von Frisch, Ludwig Maximilians University Archive, Munich, Folder E-II-1376 (Akten des Senats). In the folder at Ludwig Maximilians University is a copy of the letter to Demmel that was prepared for the university president and received by his office on May 21, 1941.

46. Von Frisch had wanted to offer Curt Stern of the Kaiser Wilhelm Institute for Biology a position as conservator, but the offer was rescinded, and Stern immigrated to the United States in 1933 and assumed a position at the University of Rochester and later at the University of California at Berkeley, where

Richard Goldschmidt had been since 1935. The letter also refers to an un-named "Mischling," whom von Frisch allegedly let stand for the *Habilitation* after 1937, and a Jewish scientific assistant named Ehrlich, whom he, according to Bergdolt, kept on at the institute as his private employee. For details on Stern, see Deichmann, *Biologists under Hitler*, 14, 21, 28, 42, 76, 385n49.

47. Karl von Frisch to Alfred Kühn, August 28, 1941, Nachlaß Karl von Frisch, ANA 540 C.II.13 "Abstammungsbescheid."

48. Arnold Sommerfeld to Ferdinand Sauerbruch, August 31, 1941, Nachlaß Karl von Frisch, ANA 540 C.II.13 "Abstammungsbescheid"; Ferdinand Sauerbruch to Arnold Sommerfeld, September 4, 1941, Nachlaß Karl von Frisch, ANA 540 C.II.13 "Abstammungsbescheid."

49. Otto Koehler to Karl von Frisch, March 3, 1942, Nachlaß Karl von Frisch, ANA 540 C.II.13 "Abstammungsbescheid."

50. Schreiber to Ferdinand Sauerbruch, October 27, 1941, and Schreiber to Gerhard Klopfer, October 27, 1941, Nachlaß Karl von Frisch, ANA 540 C.II.13 "Abstammungsbescheid."

51. Kershaw, *Hitler*, 11, 14.

52. See Sachse and Walker, "Politics and Science in Wartime."

53. For a discussion of the mutual compatibility of these cultural currents, see Herf, *Reactionary Modernism*.

54. See Müller-Hill, *Murderous Science*; Heim, "Research for Autarky"; Saraiva and Wise, "Autarky/Autarchy"; Götz and Heim, *Vordenker der Vernichtung*; and Dornheim, "Rasse, Raum und Autarkie."

55. Reinhardt, "The Entomological Institute of the Waffen-SS," 221; Jansen, *"Schädlinge"*; and Deichmann, *Biologists under Hitler*, 265 (quote).

56. For an insightful discussion of the cultural valences of the "harmers," or *Schädlinge*, up to the Weimar period, see Jansen, *"Schädlinge."*

57. Sewig, *Der Mann, der die Tiere liebte.*

58. Otto Koehler to Karl von Frisch, April 1, 1942, Nachlaß Karl von Frisch, ANA 540 C.II.13 "Abstammungsbescheid."

59. Jan Bayer to Rektor der Universität, February 26, 1942, Personalakt Karl von Frisch, Ludwig Maximilians University Archive, Munich, Folder E-II-1376 (Akten des Senats).

60. Karl von Frisch to Rektorat, March 14, 1942, Nachlaß Karl von Frisch, ANA 540 C.II.13 "Abstammungsbescheid."

61. Remarkably, Narten's talking points survive in a folder among von Frisch's papers in Munich. So seventy years later we have a very good idea of what was said behind closed doors that day. Nachlaß Karl von Frisch, ANA 540 C.II.13 "Abstammungsbescheid."

62. Reich Minister for Science, Education, and Public Instruction to Bavarian State Ministry of Education and Culture, July 27, 1942, Personalakt Karl von Frisch, Ludwig Maximilians University Archive, Munich, Folder E-II-1376 (Akten des Senats).

63. Nachlaß Karl von Frisch, ANA 540 C.II.10. Adjusted for inflation, 10,000 reichsmark converts to approximately $157,000 in 2011! He attributed his rise in income to sales of his popular biology book *Du und das Leben.*

64. Frisch, *Erinnerungen*, 113.

## Bee Vignette III

1. At the time, the films were called *Die Honigbiene I—Imkerei* (The honeybee I—apiary) and *Die Honigbiene II—Entwicklung einer Biene, Gründung eines Volkes* (The honeybee II—development of a bee, founding of a colony). The films are listed today in the IWF Wissen und Medien catalog as Karl von Frisch, *Entwicklung der Honigbiene und des Bienenvolkes* [Development of the honeybee and of a colony] (Germany: IWF, 1942–1944); Karl von Frisch, *Pollen- und Nektarsammeln der Honigbiene* [Collection of pollen and nectar in the honeybee] (Germany: IWF Wissen und Medien, 1942–1944).
2. To meet the specific needs of the schools it was meant to serve, the RWU based its film decisions largely on surveys filled out by teachers and administrators.
3. "The film 'The Honeybee' is finished but for one scene of the queen's egg laying. It's to be assumed that this last shoot will succeed in the spring of this year." Jahresbericht der Abt. IX fuer die Zeit vom 1.4.1942—31.3. 1943, 5, Bundesarchiv Berlin.
4. Von Frisch, *The Dancing Bees*, 40.
5. For especially striking examples of this, see von Frisch, *Du und das Leben* (Deutscher).
6. Von Frisch, *The Dancing Bees*, 41.
7. Rentschler, *The Ministry of Illusion*, ch. 1.
8. In the film *Der ewige Jude*, scenes with rats were included as an overt parallel to the spread and scourge of the cosmopolitan Jew. For an excellent discussion on the parallels between depictions of Jews and lice and the pesticides that were used in concentration camp gas chambers, see "Jews" in Raffles, *The Illustrated Insectopedia*, 141–61; and Russell, "'Speaking of Annihilation.'"

## Chapter Five

1. Paxton and Hessler, *Europe in the Twentieth Century*; Mazower, *Dark Continent*; Hobsbawm, *The Age of Extremes*.
2. Joseph Goebbels, "Nun, Volk, steh auf, und Sturm brich los!" Rede im Berliner Sportpalast, February 18, 1943, accessed March 16, 2015, https://archive.org/stream/WolltIhrDenTotalenKrieg/GoebbelsJoseph-Rede-WolltIhrDenTotalenKrieg194315S.#page/no/mode/1up.
3. Large, *Where Ghosts Walked*, 336–37.
4. David Anderson, "Bombs Rip Munich for 3d Day in Row," *New York Times*, July 14, 1944. The bombing raids took place on July 11, 12, and 13, 1944.
5. Ibid.
6. Alan A. Arlin, "Captain Alan A. Arlin's Diary," July 13, 1944, accessed March 16, 2015, http://www.398th.org/History/Diaries/Arlin/Arlin_440713.html. The bombing raids of mid-July—six in total—killed 3,000 people and left more than 200,000 homeless. Large, *Where Ghosts Walked*, 340.

7. Anderson, "Bombs."
8. Armand Fugge, "T/Sgt. Armand Fugge's Diary," July 16, 1944, accessed March 16, 2015, http://www.398th.org/History/Diaries/Fugge/Fugge_440716.html.
9. Von Frisch, *Erinnerungen*, 122.
10. Von Frisch, *Fünf Häuser*, 106.
11. Seeley, Kühnholz, and Seeley, "An Early Chapter."
12. Von Frisch, *Erinnerungen*, 62.
13. Von Frisch, *Fünf Häuser*, 115.
14. Seeley, Kühnholz, and Seeley, "An Early Chapter," 446.
15. Von Frisch, *Erinnerungen*, 117.
16. Ibid.
17. Von Frisch, *Fünf Häuser*, 114.
18. Ibid., 116.
19. Von Frisch, "Die Tänze der Bienen," 3. This anecdote is remembered differently in von Frisch, *The Dance Language*, 3. There he recounts that Ruth Beutler had already trained her bees to feed at a faraway location but wished them nearer to the hive. According to his telling, the bees continued to show up preferentially at the more distant site rather than the nearby spot, suggesting to him that the dances contained distance information. Common to both accounts is the significance of the interaction with Ruth Beutler, which seems to have prompted him to revisit the bee dances. Here I have chosen to narrate the recollection that was published nearer to the alleged events, with the caveat that the exact nature of the interaction between them is unknown and likely not recoverable.
20. We are afforded an especially candid glimpse of these developments through a set of unpublished laboratory notebooks that survive at the state library in Munich: Nachlaß Karl von Frisch, Bayerische Staatsbibliothek, Munich, ANA 540 A.III. On von Frisch's notebooks, see also Hoffmann, "Retenir et Reprendre."
21. Von Frisch, "Die Tänze der Bienen," 5–6.
22. Von Frisch, "Die Werbetänze der Bienen."
23. Von Frisch, "Die Tänze der Bienen," 7.
24. Karl von Frisch, Versuchsprotokolle, III, August 12, 1944, Nachlaß Karl von Frisch, Bayerische Staatsbibliothek, Munich, ANA 540 A.III.
25. Von Frisch, "Die Tänze der Bienen," 27.
26. Karl von Frisch, Versuchsprotokolle, III, August 17–18, 1944, Nachlaß Karl von Frisch, Bayerische Staatsbibliothek, Munich, ANA 540 A.III.
27. Named after the nineteenth-century scientist C. Lloyd Morgan, Morgan's canon stipulates that whenever a simple explanation will do to account for a given behavior, then a more complex account should be avoided. Richards, *Darwin and the Emergence of Evolutionary Theories*; Radick, "Morgan's Canon."
28. Von Frisch, *Erinnerungen*, 127.
29. Von Frisch, *Fünf Häuser*, 110.
30. Ibid., 111–12.
31. Ibid., 111.

## Chapter Six

1. Likens, "Arthur David Hasler," 3.
2. Gimbel, *Science, Technology, and Reparations*, chs. 1–2.
3. Large, *Where Ghosts Walked*, 338, 346 (quote).
4. Hasler, "A War Time View," 82.
5. Ibid., 83.
6. Von Frisch, "Die Tänze der Bienen," 18.
7. Ibid., 18.
8. Hasler, "This Is the Enemy."
9. Von Frisch, *Fünf Häuser*, 117.
10. Ibid.
11. Von Frisch, "Die Tänze der Bienen," 34.
12. Ibid., 33, 36 (quote).
13. For example, Santschi, "Observations."
14. Brun, *Raumorientierung*; Lubbock, *Ants, Bees, and Wasps*; Wheeler, *Ants*; Forel, *Les Fourmis de la Suisse*.
15. Wolf, "Über das Heimkehrvermögen der Bienen," 234.
16. Karl von Frisch to Otto Koehler, January 12, 1946, Nachlaß Karl von Frisch, Bayerische Staatsbibliothek, Munich, ANA 540 B.II.
17. Henkel, "Unterscheiden die Bienen Tänze?"; and von Frisch, "Die Werbetänze," 271.
18. Von Frisch, "Die 'Sprache' der Bienen," 400n.
19. Von Frisch, "Die Tänze der Bienen," 3.
20. Von Frisch, *Über die "Sprache" der Bienen*, 44.
21. Von Frisch, "Die Tänze der Bienen," 26.
22. Their reasons for abandoning the distinction between the dances are twofold: First, they argue that the transitions between the two shapes are continuous rather than abrupt and therefore no clear separation between them can be identified. Second, they have found that even the dances for near sources contain usable information about direction. And yet the distinction between the two dances nonetheless persists in many textbooks on animal behavior and bee biology to this day. See Gardner, Seeley, and Calderone, "Do Honeybees"; Griffin, Smith, and Seeley, "Do Honeybees." I thank Francis Ratnieks and Thomas Seeley for this information.
23. Arthur Hasler to George H. Parker and Dwight Minnich, August 25, 1945, Rockefeller Foundation Archive, Sleepy Hollow, New York (hereafter designated RFA), RG 1.1, Series 717D, Box 14, Folder 136 (University of Munich 1948–1952).
24. Hasler, "This Is the Enemy."
25. Hasler, "A War Time View," 83.
26. Hasler to George H. Parker and Dwight Minnich, August 25, 1945, RFA, RG 1.1, Series 717D, Box 14, Folder 136 (University of Munich 1948–1952).
27. Arthur Hasler to Fosdick, August 26, 1945, RFA, RG 1.1, Series 717D, Box 14, Folder 136 (University of Munich 1948–1952).

28. Hasler, "This Is the Enemy."
29. Appleget to Hasler, September 25, 1945, RFA, RG 1.1, Series 717D, Box 14, Folder 136 (University of Munich 1948–1952).
30. Wiecki, "Professors in Purgatory," 85n.
31. Tent, *Mission on the Rhine*, 83.
32. Wiecki, "Professors in Purgatory," 172.
33. Oeffentliche Kläger Heine to Karl von Frisch (copy), May 5, 1950, Bayerisches Hauptstaatsarchiv, Munich, MK ad54482.
34. Ziemke, *The US Army and the Occupation of Germany*, 430n.
35. Wiecki, "Professors in Purgatory," 113–14.
36. Patton quoted in Raymond Daniell, "Patton Belittles Denazificiation; Holds Rebuilding More Important," *New York Times*, September 23, 1945; and Wiecki, "Professors in Purgatory," 93–95.
37. Wiecki, "Professors in Purgatory," 179–80.
38. The statistics are from ibid., 174–75. See also Huber, "Die Universität München"; and Boehling, *A Question of Priorities*.
39. Deichmann, *Biologists under Hitler*, 298–317.
40. On the halting postwar reintegration of German biologists (especially geneticists) and Niko Tinbergen's intial aversion to German scientists based on his wartime experiences, respectively, see Deichmann, *Biologists under Hitler*, 305–14; and Burkhardt, *Patterns*, 283–89.
41. Sachse, "What Research, to What End?," 120.
42. Von Frisch, *Erinnerungen*, 131, 133.
43. Ibid., 132–33.
44. Part 1 was published in German as "Gelöste und ungelöste Rätsel der Bienensprache, I" and part 2 as "Gelöste und ungelöste Rätsel der Bienensprache, II."
45. Von Frisch, *Bees*, 86.
46. Von Frisch, "Die Polarisation," 143.
47. Von Frisch, *Bees*, 90.
48. Von Frisch, "Die Polarisation," 145n.
49. Von Frisch, "Dances of the Honey Bee."
50. Thorpe, "Orientation and Methods," 11.
51. For a compromise in which the dances are a first among several factors in bee recruitment, see Tautz, *Die Erforschung der Bienenwelt*, 58–75.
52. Thorpe, "Orientation and Methods," 14.
53. Pauly, *Controlling Life*.
54. Thorpe, "Orientation and Methods," 11–12.
55. Sachse, "What Research, to What End?," 113.
56. Robert Havinghurst, diary entry for September 10, 1948, Zurich, RFA, RG 1.2, Series 705D, Box 6, Folder 56.
57. [Donald R. Griffin], "Cornell University Proposal for Travel Grant," November 5, 1948, RFA, RG 1.2, Series 705D, Box 6, Folder 56.
58. J. H. Welsh to Donald Griffin, October 4, 1948. This document is included in the application to the Rockefeller Foundation cited above.

## Chapter Seven

1. Donald R. Griffin to Dear Sir, March 24, 1949, Rockefeller Foundation Archive, Sleepy Hollow, New York (hereafter designated RFA), RG 1.2, Series 705D, Box 6, Folder 56.
2. Ibid.
3. The discussion of von Frisch's lecture series in based on the published version of the talks: von Frisch, *Bees*, esp. 35 (quote), 48, and 85 (quote).
4. See Dyer and Seeley, "Dance Dialects"; and Gould, "Why do Honey Bees Have Dialects?"
5. Watson, *Behaviorism*, 82.
6. On American laboratory research of behavior, especially at Yale University, see Lemov, *World as Laboratory*.
7. Beach, "The Snark Was a Boojum," 116, 119.
8. Von Frisch, *Erinnerungen*, 143.
9. The experiments were performed by Robert Williams, an endocrinologist at Harvard. Karl von Frisch, diary entry for April 2, 1949, Nachlaß Karl von Frisch, Bayerische Staatsbibliothek, Munich, ANA 540 C.I (hereafter designated Nachlaß Karl von Frisch).
10. Von Frisch, *A Biologist*, 158–59.
11. Bonanos, *Instant*, 19–20.
12. Karl von Frisch, diary entry for April 2, 1949, Nachlaß Karl von Frisch, ANA 540 C.I.
13. Bonanos, *Instant*, 44.
14. Robert Plumb, "Scientist Learns 'Dance' of Bees, Sends Workers 2 Miles for Nectar," *New York Times*, March 16, 1949.
15. "Dance of Bees Tells Way to Food Cache," *New York Times*, April 7, 1949.
16. Von Frisch, *A Biologist*, 144.
17. Von Frisch, *Erinnerungen*, 144.
18. Karl von Frisch, diary entry for April 8, 1949, Nachlaß Karl von Frisch, ANA 540 C.I. The summary of von Frisch's Rockefeller Foundation visit is based on this entry.
19. Walker, *Nazi Science*, ch. 2.
20. Von Frisch, *Erinnerungen*, 145; and Karl von Frisch, diary entry for April 13, 1949, Nachlaß Karl von Frisch, ANA 540 C.I.
21. June Owen, "News of Food: Milk: Walker-Gordon Farm Specializes in a Product of Extra-High Quality," *New York Times*, September 5, 1958, 20.
22. Karl von Frisch, diary entry for April 12, 1949, Nachlaß Karl von Frisch, ANA 540 C.I.
23. Karl von Frisch, diary entries for April 17–23, 1949, Nachlaß Karl von Frisch, ANA 540 C.I.
24. Hayes, *The Ape in Our House*, 7.
25. Von Frisch, *Erinnerungen*, 145.
26. Karl von Frisch, diary entry for April 17, 1949, Nachlaß Karl von Frisch, ANA 540 C.I (emphasis von Frisch).
27. Hayes, *The Ape in Our House*, 1, 9.

28. Karl von Frisch, diary entry for April 23, 1949, Nachlaß Karl von Frisch, ANA 540 C.I.
29. Von Frisch, *Erinnerungen*, 150.
30. Ibid., 153.
31. Griffin, foreword to von Frisch, *Bees*, vii–viii.
32. Karl von Frisch, diary entry for April 18, 1949, Nachlaß Karl von Frisch, ANA 540 C.I.
33. Griffin, foreword to von Frisch, *Bees*, ix; Donald R. Griffin to Paine, April 1, 1949, RFA, RG 1.2, Series 705D, Box 6, Folder 56.
34. Griffin, foreword, to von Frisch, *Bees*, vii.
35. Robert Yerkes to Donald R. Griffin, March 30, 1949, RFA, RG 1.2, Series 705D, Box 6, Folder 56.
36. Karl von Frisch, diary entry for April 8, 1949, Nachlaß Karl von Frisch, ANA 540 C.I. On American funding of European research in the postwar period, see Krige, *American Hegemony*.
37. Exner, *Physiology of the Compound Eyes*, vii.
38. Von Frisch, "Die Sonne als Kompass," 212.
39. Ibid.
40. Hasler, foreword to Hasler and Scholz, *Olfactory Imprinting*, xii.
41. Ibid.
42. The work Hasler reviewed for *Science* in 1948 was von Frisch's *Duftgelenkte Bienen im Dienste der Landwirtschaft und Imkerei*.
43. Von Frisch, *Bees*, xii.
44. When von Frisch left Graz again for Munich in early 1950, the Rockefeller Foundation reduced his annual award to $2,500 but would later raise it again. RFA, RG 1.2, Series 705D, Box 6, Folder 56. Karl von Frisch, July 3, 1964, RFA, RG 1.1, Series 717D, Box 15, Folder 139.

## Bee Vignette IV

1. Aristotle, *Historia animalium*, bk. 9, ch. 40.
2. Historians of science have significantly complicated the concept of the Scientific Revolution. For very accessible introductions to both the period and scholarship that has qualified the concept, see Dear, *Revolutionizing the Sciences*; Shapin, *The Scientific Revolution*; and Sivin, "Why Did the Scientific Revolution Not Take Place in China—or Didn't It?"
3. Daston, "The Empire of Observation," 81–115 (quoted material from p. 93).
4. Jardine, *Ingenious Pursuits*, esp. chs. 2 and 3.
5. Ibid., 103.
6. Ruestow, *The Microscope in the Dutch Republic*.
7. Swammerdam, *Book of Nature*, 106.
8. Swammerdam quoted in Mazzolini, "Schirach's Experiments on Bees," 71, 74.
9. On collaboration between artists and microscopists in general, see Jardine, *Ingenious Pursuits*, 101.
10. Swammerdam, *Book of Nature*, 187.

11. The subsequent discussion of Huber's scientific work is based on Huber, *Observations upon Bees*. Quoted material is from pages 17, 624, 3, 3–4, 4, 20–21, 21, 27, 28, 29, and 31, respectively. I thank Michael Bush for his beautiful work on this book.

12. Von Frisch, *Über die "Sprache" der Bienen*, 19.

13. Burkhardt, *Patterns*, esp. the introduction and chs. 1 and 2.

14. Von Frisch, *Aus dem Leben der Bienen* (1931), v.

15. Gries and Koeniger, "Straight Forward to the Queen," 541.

16. On drones as weapons, see Zerner, "Stealth Nature."

### Chapter Eight

1. Klaus Köhne, "Nobelpreis kam beim Mittagsschlaf: Professor von Frisch: 'Damit haette ich nie gerechnet,'" *Münchner Stadtzeitung*, October 12, 1973, 1.

2. Burkhardt, *Patterns*, chs. 3–9.

3. Börje Cronholm, "The Nobel Prize in Physiology or Medicine 1973, Award Ceremony Speech," accessed March 16, 2015, http://www.nobelprize.org/nobel_prizes/medicine/laureates/1973/presentation-speech.html.

4. On the history of the maternal instinct, see Vicedo, *The Nature and Nurture of Love*.

5. Cronholm, "The Nobel Prize."

6. Wenner, "Honey Bees" (1967); Johnson, "Honey Bees." In 1967, when Wenner's first *Science* article appeared, von Frisch was already eighty-one years old. This was three years after he had performed his last bee experiments. Karl von Frisch to Otto Koehler, October 16, 1965, Nachlaß Karl von Frisch, Bayerische Staatsbibliothek, Munich, ANA 540 B.II (hereafter designated Nachlaß Karl von Frisch).

7. Wenner, Wells, and Johnson, "Honeybee Recruitment." See also Munz, "The Bee Battles;" Gould, "Honey Bee Recruitment"; Veldink, "Paradigm Challenges in Modern Science," ch. 2; Veldink, "The Honey-Bee Language Controversy"; Wenner and Wells, *Anatomy of a Controversy*; and Kak, "The Honey Bee Dance Language Controversy."

8. For a good introduction to behaviorism and its rise, see Boakes, *From Darwin to Behaviorism*, esp. ch. 6.

9. For an overview of these developments, see Smith, *Norton History of the Human Sciences*; Barsky, *Noam Chomsky*; Harris, *The Linguistics Wars*; Cohen-Cole, *The Open Mind*; Greenwood, "Understanding the 'Cognitive Revolution'"; Mandler, "Origins of the Cognitive Revolution"; and Dewsbury, *Comparative Psychology in the Twentieth Century*.

10. Stuart Altman, "'Without a Word'" 361.

11. Von Frisch, *Aus dem Leben* (1927), 116. The seventh edition was published in 1964.

12. Griffin, *The Question of Animal Awareness*, 16–17; and on bee language, see ibid., 19–25, 92–94.

13. Karl von Frisch to Donald Griffin, November 13, 1976, Nachlaß Karl von Frisch, ANA 540 B.II.

14. Ibid.
15. See, for example, Hockett, *A Course in Modern Linguistics*, ch. 64; Lotz, "Speech and Language"; and Lotz, "Book Review."
16. Révész, "Der Kampf," 83. On affective versus social interpretations of animal vocalizations, see Montgomery, "Place, Practice and Primatology"; and Thomas, "Are Animals Just Noisy Machines?"
17. Von Frisch, "'Sprache' oder 'Kommunikation'"; and Von Frisch, *Tanzsprache und Orientierung.*
18. Von Frisch had already suspected that the dance language was used by bees to communicate possible new hive sites during the swarming period. Von Frisch, "Die 'Sprache' der Bienen."
19. See Lindauer, "Schwarmbienen auf Wohnungssuche"; and Lindauer, *Communication among Social Bees.*
20. Von Frisch, "'Sprache' oder 'Kommunikation,'" 236. On von Frisch's use of quotation marks, see also von Frisch, *Tanzsprache und Orientierung*, 1.
21. Von Frisch continued to perform variations of these experiments into the early 1960s. See von Frisch, *Tanzsprache und Orientierung*, 85–96, 156–63.
22. Wenner, "Honey Bees" (1967), 849; Johnson, "Honey Bees," 847.
23. Von Frisch, Wenner, and Johnson, "Honeybees."
24. Von Frisch, *Tanzsprache und Orientierung*, 177–91.
25. Von Frisch, Wenner, and Johnson, "Honeybees," 1075, 1073, respectively. Wenner himself later abandoned the idea that pheromones from the Nasonov glands serve to attract other bees to food sources.
26. Ibid., 1077, 1076, respectively.
27. Wenner, Wells, and Johnson, "An Analysis," 326.
28. Karl von Frisch to Rudolf Jander, September 11, 1963, Nachlaß Karl von Frisch, ANA 540 B.II.
29. Karl von Frisch to Rudolf Jander, January 19, 1968, Nachlaß Karl von Frisch, ANA 540 B.II.
30. Wenner, "Information Transfer in Honey Bees," 134–35.
31. Wenner and Wells, *Anatomy of a Controversy*, 362–66.
32. Adrian Wenner, "Read Me First—a Chronology," accessed March 16, 2015, http://www.beesource.com/point-of-view/adrian-wenner-read-me-first/. Emphasis in original.
33. Karl von Frisch to James L. Gould, April 19, 1975, Nachlaß Karl von Frisch, ANA 540 B.II.
34. Gould, Henerey, and MacLeod, "Communication of Direction," 553.
35. Donald Griffin to Karl von Frisch, November 25, 1974, Nachlaß Karl von Frisch, ANA 540 B.IV.
36. Ernst Mayr to Karl von Frisch, May 25, 1967, Nachlaß Karl von Frisch, ANA 540 B.IV.
37. See, for example, Karl von Frisch to Heini Hediger, July 19, 1973, Nachlaß Karl von Frisch, ANA 540 B.II; and Karl von Frisch to Martin Lindauer, February 18, 1964, and May 3, 1974, Nachlaß Karl von Frisch, ANA 540 B.II.
38. Thorpe, *The Origins and Rise of Ethology*, 65.
39. For a more in-depth discussion of Wenner's challenge and Gould's work with a similar upshot, see Manning, *An Introduction to Animal Behaviour*, 107.

40. Griffin, *The Question of Animal Awareness*, 27 and 29.
41. Wilson, "Letters," 6.
42. Altman, "'Without a Word,'" 361, quoted in Davenport, "'Bee Language.'"
43. Wenner and Wells, *Anatomy of a Controversy*, 200–204, 274–84. See also Wells and Wenner, "Do Honey Bees Have a Language?"; and Wenner, "Information Transfer in Honey Bees."
44. Gould, "Honey Bee Recruitment," 692. See also Crist, "Can an Insect Speak?"

## Conclusion

1. Karl von Frisch, diary entries from May 2, 1980, to June 5, 1982, Nachlaß Karl von Frisch, Bayerische Staatsbibliothek, Munich, ANA 540 C.I.
2. Seeley, *Honeybee Democracy*, 218, 230.
3. Ibid., 5, 220.
4. Kosek, "Ecologics of Empire"; Kosek, *Homo Apians*.
5. United States Department of Agriculture, *Colony Collapse Disorder Progress Report*, June 2010, p. 6, accessed March 16, 2015, www.ars.usda.gov/is/br/ccd/ccdprogressreport2010.pdf.
6. Edward O. Wilson on Thomas Seeley's *Honeybee Democracy*, accessed March 16, 2015, http://press.princeton.edu/quotes/q9267.html.

# Bibliography

## Archives and Collections Consulted

American Philosophical Society, Philadelphia, PA
Bayerische Akademie der Wissenschaften Archive, Munich, Germany
Bayerisches Hauptstaatsarchiv, Munich, Germany
    *Personalakt Karl von Frisch*
    *Institusakten des Zoologischen Instituts*
Bayerische Staatsbibliothek, Munich, Germany
    *Nachlaß Karl von Frisch, ANA 540*
Bundesarchiv Berlin, Berlin, Germany
    *Personalakten Karl von Frisch*
Ludwig Maximilians University Archive, Munich, Germany
    *Personalakt Karl von Frisch*
National Archives and Records Administration, Washington, DC
Rockefeller Foundation Archive, Sleepy Hollow, NY
Staatsbibliothek zu Berlin (West), Preußischer Kulturbesitz, Berlin, Germany
University of Vienna Archive, Vienna, Austria

## Published Sources

Ackerl, Isabella. *Vienna Modernism*. Vienna: Federal Press Service, 1999.

Adams, Mark A., ed. *The Wellborn Science: Eugenics in Germany, France, Brazil, and Russia*. Monographs on the History and Philosophy of Biology. New York: Oxford University Press, 1990.

Allen, Danielle. "Burning the Fable of the Bees: The Incendiary Authority of Nature." In *The Moral Authority of Nature*, edited by Loraine Daston and Fernando Vidal, 74–99. Chicago: University of Chicago Press, 2004.

Altman, Stuart A. "Without a Word." Review of *Non-Verbal Communication*, edited by Robert A. Hinde. *Nature* 240 (1972): 361.

Aristotle. *Historia animalium*. Loeb Classical Library. London: Heinemann, 1965.

Azzouni, Safia. "Wissenschaftspopularisierung um 1900 als exemplarisch-literarische Rekonstruktion bei Wilhelm Bölsche." In *Das Beispiel: Epistemologie des Exemplarischen*, edited by Nicolas Pethes, Jens Ruchatz, and Stefan Willer, 278–93. Berlin: Kadmos, 2007.

Barsky, Robert F. *Noam Chomsky: A Life of Dissent*. Cambridge, MA: MIT Press, 1997.

Bashford, Alison, and Philippa Levine, eds. *The Oxford Handbook of the History of Eugenics*. New York: Oxford University Press, 2010.

Beach, Frank A. "The Snark Was a Boojum." *American Psychologist* 5 (1950): 115–24.

Benz, Wolfgang. *Die Juden in Deutschland, 1933–1945: Leben unter Nationalsozialistischer Herrschaft*. Munich: Beck, 1988.

Beyerchen, Alan. *Scientists under Hitler: Politics and the Physics Community in the Third Reich*. New Haven, CT: Yale University Press, 1977.

Boakes, Robert. *From Darwin to Behaviourism: Psychology and the Minds of Animals*. Cambridge: Cambridge University Press, 1984.

Boehling, Rebecca. *A Question of Priorities: Democratic Reforms and Economic Recovery in Postwar Germany*. Providence, RI: Berghahn Books, 1996.

Böhm, Helmut. *Von der Selbstverwaltung zum Führerprinzip: Die Universität München in den ersten Jahren des Dritten Reiches (1933–1936)*. Berlin: Duncker and Humblot, 1995.

Bonanos, Christopher. *Instant: The Story of Polaroid*. New York: Princeton Architectural Press, 2012.

Bracher, Karl Dietrich. "Stages of Totalitarian 'Integration' (*Gleichschaltung*): The Consolidation of National Socialist Rule in 1933 and 1934." In *Republic to Reich: The Making of the Nazi Revolution; Ten Essays*, edited by Hajo Holborn, 109–28. New York: Pantheon Books, 1972.

Brauckmann, Sabine, and Gerd Müller, eds. *A Laboratory in the Prater: Experiment and Theory in the Biologische Versuchsanstalt in Vienna*. Cambridge, MA: MIT Press, forthcoming.

Braun, Marta, and Etienne-Jules Marey. *Picturing Time: The Work of Etienne-Jules Marey (1830–1904)*. Chicago: University of Chicago Press, 1992.

Brun, Rudolf. *Die Raumorientierung der Ameisen und das Orientierungsproblem im allgemeinen: Eine kritisch-experimentelle Studie; zugleich ein Beitrag zur Theorie der Mneme*. Jena: Gustav Fischer, 1914.

Burkhardt, Richard W. "Karl von Frisch." In *Dictionary of Scientific Biography*, vol. 17, supplement 2, 313–20. New York: Scribner, 1990.

———. *Patterns of Behavior: Konrad Lorenz, Niko Tinbergen, and the Founding of Ethology*. Chicago: University of Chicago Press, 2005.

Burleigh, Michael. *The Third Reich: A New History*. New York: Hill and Wang, 2000.

Burleigh, Michael, and Wolfgang Wippermann. *The Racial State: Germany, 1933–1945*. Cambridge: Cambridge University Press, 1991.

Burt, Jonathan. *Animals in Film*. London: Reaktion, 2002.

Busch, Wilhelm. *Schnurrdiburr, oder: Die Bienen*. 6th ed. Munich: Braun und Schneider, 1884.

Clark, John. "'The Complete Biography of Every Animal': Ants, Bees, and Humanity in Nineteenth-Century England." *Studies in History and Philosophy of Biological and Biomedical Sciences* 29 (1998): 249–67.

Coen, Deborah R. "A Lens of Many Facets: Science through a Family's Eyes." *Isis* 97 (2006): 395–419.

———. "Living Precisely in Fin-de-Siècle Vienna." *Journal of the History of Biology* 39 (2006): 493–523.

———. *Vienna in the Age of Uncertainty: Science, Liberalism, and Private Life*. Chicago: University of Chicago Press, 2007.

Cohen-Cole, Jamie. *The Open Mind: Cold War Politics and the Sciences of Human Nature*. Chicago: University of Chicago Press, 2014.

Corni, Gustavo, and Horst Gies. *Brot, Butter, Kanonen: Die Ernährungswirtschaft in Deutschland unter der Diktatur Hitlers*. Berlin: Akademie Verlag, 1997.

Crist, Eileen. "Can an Insect Speak? The Case of the Honeybee Dance Language." *Social Studies of Science* 34 (2004): 7–43.

Dagognet, François. *Etienne-Jules Marey: A Passion for the Trace*. New York: Zone Books, 1992.

Danto, Arthur Coleman. "Giotto and the Stench of Lazarus." In *Philosophizing Art: Selected Essays*, 106–22. Berkeley: University of California Press, 1999.

Darwin, Charles. *The Descent of Man, and Selection in Relation to Sex*. Princeton, NJ: Princeton University Press, 1981.

———. *On the Origin of Species: A Facsimile of the First Edition*. Cambridge, MA: Harvard University Press, 2001.

Daston, Lorraine. "The Empire of Observation, 1600–1800." In *Histories of Scientific Observation*, edited by Lorraine Daston and Elizabeth Lunbeck, 81–114. Chicago: University of Chicago Press, 2011.

Daston, Lorraine, and Peter Galison. "The Image of Objectivity." *Representations* 40, 1 (1992): 81–128.

Daston, Lorraine, and Elizabeth Lunbeck, eds. *Histories of Scientific Observation*. Chicago: University of Chicago Press, 2011.

Daston, Lorraine, and Fernando Vidal, eds. *The Moral Authority of Nature*. Chicago: University of Chicago Press, 2004.

Daum, Andreas. *Wissenschaftspopularisierung im 19. Jahrhundert: Bürgerliche Kultur, naturwissenschaftliche Bildung und die deutsche Öffentlichkeit 1848–1914*. Munich: Oldenbourg, 2002.

Davenport, Demorest. "'Bee Language' Letters." *Science* 186 (1974): 975.

Dawkins, Richard. "'Bees Are Easily Distracted' Letters." *Science* 165 (1969): 751.

Dear, Peter. *Revolutionizing the Sciences*. Princeton, NJ: Princeton University Press, 2009.

Deichmann, Ute. *Biologen unter Hitler: Porträt einer Wissenschaft im NS-Staat*. Geschichte Fischer. Frankfurt am Main: Fischer Taschenbuch Verlag, 1995.

———. *Biologists under Hitler*. Translated by Thomas Dunlap. Cambridge, MA: Harvard University Press, 1996.

Dewsbury, Donald. *Comparative Psychology in the Twentieth Century*. Stroudsburg, PA: Reinhold, 1984.

Doflein, Franz. "Der angebliche Farbensinn der Insekten." *Die Naturwissenschaften* 29 (1914): 708–10.

Dornheim, Andreas. "Rasse, Raum und Autarkie: Sachverständigengutachten zur Rolle des Reichsministeriums für Ernährung und Landwirtschaft in der NS-Zeit." Bundesministerium für Ernährung, Landwirtschaft und Verbraucherschutz, Berlin, 2011.

Drouin, Jean-Marc. "L'Image des Sociétés d'Insectes en France à l'Epoque de la Revolution." *Revue de Synthèse* 4 (1992): 333–45.

Dyer, Fred C., and Thomas D. Seeley. "Dance Dialects and Foraging Range in Three Asian Honey Bee Species." *Behavioral Ecology and Sociology* 28, 4 (1999): 227–33.

Elon, Amos. *The Pity of It All: A History of Jews in Germany, 1743–1933*. New York: Metropolitan Books, 2002.

Evans, Richard J. *The Third Reich in Power, 1933–1939*. New York: Penguin Press, 2005.

Exner, Franz, and Sigmund Exner. "Die physikalischen Grundlagen der Blütenfärbungen." *Sitzungsberichte der Mathematisch-Naturwissenschaftlichen Klasse der Kaiserlichen Akademie der Wissenschaften*, 1910, 191–245.

Exner, Sigmund. *The Physiology of the Compound Eyes of Insects and Crustaceans.* Translated by Roger C. Hardie. With an introduction by Karl von Frisch. Berlin: Springer Verlag, 1989.

Falk, Raphael. *Genetic Analysis: A History of Genetic Thinking*. Cambridge Studies in Philosophy and Biology. Cambridge: Cambridge University Press, 2009.

Forel, Auguste. *Les Fourmis de la Suisse*. 1874. La Chaux-de Fonds: Imprimerie Coopérative, 1924.

Frisch, Karl von. *Aus dem Leben der Bienen*. Berlin: J. Springer, 1927.

———. *Aus dem Leben der Bienen*. 2nd ed. Berlin: J. Springer, 1931.

———. *Aus dem Leben der Bienen*. 7th ed. Berlin: Springer, 1964.

———. *Bees: Their Vision, Chemical Senses, and Language*. Ithaca, NY: Cornell University Press, 1950.

———. "Beiträge zur Physiologie der Pigmentzellen in der Fischhaut." In *Festschrift zum sechzigsten Geburtstag Richard Hertwigs*, vol. 3, 15–28. Jena: G. Fischer, 1910.

———. "Beiträge zur Physiologie der Pigmentzellen in der Fischhaut." *Pflügers Archiv für Physiologie* 138 (1911): 319–87.

———. "Bericht über die am Münchner Zoologischen Institut eingeleiteten Nosemaarbeiten zur Frage der Vorbeugungsmittel, der chemotherapeutischen Bekämpfung und des Beobachtungsdienstes." *Deutscher Imkerführer* 16, 12 (1943).

———. *A Biologist Remembers*. Translated by Lisbeth Gombrich. Oxford: Pergamon Press, 1967.

———. *The Dance Language and Orientation of Bees*. Cambridge, MA: Harvard University Press, 1993.

———. "Dances of the Honey Bee." *Bulletin of Animal Behaviour* 5 (1947): 1–32.

———. *The Dancing Bees: An Account of the Life and Senses of the Honey Bee*. 2nd ed. London: Methuen, 1966.

———. *Du und das Leben*. Berlin: Deutscher Verlag, Dr.-Goebbels-Spende für die deutsche Wehrmacht, 1936.

———. *Du und das Leben: Eine Moderne Biologie für Jedermann*. Berlin: Ullstein Verlag, 1936.

———. "Erinnerungen an Otto Koehler aus alter Zeit." *Zeitschrift für Tierpsychologie* 35 (1974): 465–67.

———. *Erinnerungen eines Biologen*. Berlin: Springer, 1957.

————. *Der Farbensinn und Formensinn der Biene.* Jena: Gustav Fischer, 1914.

————. *Fünf Häuser am See: Der Brunnwinkl, Werden und Wesen eines Sommer-sitzes.* Berlin: Springer Verlag, 1980.

————. "Gelöste und ungelöste Rätsel der Bienensprache, I." *Die Naturwissen-schaften* 35, 1 (1948): 12–23.

————. "Gelöste und ungelöste Rätsel der Bienensprache, II." *Die Naturwissen-schaften* 35, 2 (1948): 38–43.

————. "Die Polarisation des Himmelslichtes als orientierender Faktor bei den Tänzen der Bienen." *Experientia* 5, 4 (1949): 142–48.

————. "Psychologie der Bienen." *Zeitschrift für Tierpsychologie* 1 (1937): 9–21.

————. *Sechs Vorträge über Bakteriologie für Krankenschwestern.* Vienna: Hölder Verlag, 1918.

————. "Sind die Fische Farbenblind?" *Zoologische Jahrbücher* 33 (1912): 107–26.

————. "Die Sonne als Kompass im Leben der Bienen." *Experientia* 6 (1950): 210–21.

————. "Die 'Sprache' der Bienen und ihre Nutzanwendung in der Land-wirtschaft." *Experientia* 2, 10 (1946): 397–404.

————. "'Sprache' oder 'Kommunikation' der Bienen?" *Psychologische Rund-schau* 4 (1953): 235–36.

————. "Studien über Pigmentverschiebung im Facettenauge." *Biologisches Zentralblatt* 28 (1908): 662–71, 698–704.

————. "Die Tänze der Bienen." *Österreichische Zoologische Zeitschrift* 1 (1946): 1–48.

————. *Tanzsprache und Orientierung der Bienen.* Berlin: Springer, 1965.

————. "Über den Farbensinn der Bienen und die Blumenfarben." *Muenchener Medizinische Wochenschrift* 1 (1913): 15–18.

————. "Über den Farbensinn der Fische." *Verhandlungen der Deutschen Zoolo-gischen Gesellschaft* 21 (1911): 220–25.

————. "Über den Gehörsinn der Fische." *Biological Reviews* 11, 2 (1936): 210–46.

————. "Über den Geruchssinn der Bienen und seine blütenbiologische Bedeu-tung." *Zoologische Jahrbücher* 37 (1919): 1–238.

————. "Über die 'Sprache' der Bienen, I. Mitteilung." *Münchener Medizinische Wochenschrift* 20 (1920): 566–69.

————. "Über die 'Sprache' der Bienen, II. Mitteilung." *Münchener Medizinische Wochenschrift* 17 (1921): 509–11.

————. "Über die 'Sprache' der Bienen, III. Mitteilung." *Münchener Medizinische Wochenschrift* 21 (1922): 781–82.

————. *Über die "Sprache" der Bienen: Eine tierpsychologische Untersuchung.* Jena: G. Fischer, 1923.

————. "Über farbige Anpassung bei Fischen." *Zoologische Jahrbücher* 32 (1912): 171–230.

————. "Weitere Untersuchungen über den Farbensinn der Fische." *Zoologische Jahrbücher* 34 (1913): 43–68.

————. "Die Werbetänze der Bienen und ihre Auslösung." *Naturwissenschaften* 30, 19 (1942): 269–77.

————. "Ein Zwergwels der kommt, wenn man ihm pfeift." *Biologisches Zentral-blatt* 43 (1923): 439–46.

Von Frisch, Karl, and Theodor Kollmann. *Der Neubau des Zoologischen Instituts der Universität München*. Munich: A. Huber, 1935.

Von Frisch, Karl, Adrian Wenner, and Dennis L. Johnson. "Honeybees: Do They Use Direction and Distance Information Provided by Their Dancers?" *Science* 158, 3804 (1967): 1072–77.

Gardner, Kathryn E., Thomas D. Seeley, and Nicholas W. Calderone. "Do Honeybees Have Two Discrete Dances to Advertise Food Sources?" *Animal Behaviour* 75 (2008): 1291–1300.

Gellately, Robert. *The Gestapo and German Society: Enforcing Racial Policy, 1933–45*. Oxford: Oxford University Press, 1990.

Gemelli, Giuliana, and Roy MacLeod, eds. *American Foundations in Europe: Grant-Giving Policies, Cultural Diplomacy, and Trans-Atlantic Relations, 1920–1980*. Brussels: Peter Lang, 2003.

Gemelli, Giuliana, Jean-François Picard, and William H. Schneider, eds. *Managing Medical Research in Europe: The Role of the Rockefeller Foundation (1920s–1950s)*. Bologna: CLUEB, 1999.

Giese, Clemens, and Alfred Zschiesche. *Die deutsche Tierschutzgesetzgebung*. Leipzig: Barth, 1938.

Gimbel, John. *Science, Technology, and Reparations: Exploitation and Plunder in Postwar Germany*. Stanford, CA: Stanford University Press, 1990.

Götz, Aly, and Susanne Heim. *Vordenker der Vernichtung: Auschwitz und die deutschen Pläne für eine neue europäische Ordnung*. 5th ed. Hamburg: Fischer Verlag, 2004.

Gould, James L. "Honey Bee Recruitment: The Dance-Language Controversy." *Science* 189 (1975): 685–93.

———. "Why Do Honey Bees Have Dialects?" *Behavioral Ecology and Sociobiology* 10, 1 (1982): 53–56.

Gould, James L., Michael Henerey, and Michael MacLeod. "Communication of Direction by the Honey Bee." *Science* 169 (1970): 544–54.

Greenwood, John D. "Understanding the 'Cognitive Revolution' in Psychology." *Journal of the History of the Behavioral Sciences* 35 (1999): 1–22.

Gries, Michael, and Nikolaus Koeniger. "Straight Forward to the Queen: Pursuing Honeybee Drones (*Apis mellifera L.*) Adjust Their Body Axis to the Direction of the Queen." *Journal for Comparative Physiolgoy* 179 (1996): 539–44.

Griffin, Donald R. Foreword to *Bees: Their Vision, Chemical Senses, and Language*, by Karl von Frisch, vii–xiii. Ithaca, NY: Cornell University Press, 1950.

———. *The Question of Animal Awareness: Evolutionary Continuity of Mental Experience*. New York: Rockefeller University Press, 1976.

Griffin, Donald R., and Peter Marler, "'Scientific Methods in Ethology,' Letters." *Science* 185 (1974): 814.

Griffin, Sean R., Michael L. Smith, and Thomas D. Seeley. "Do Honeybees Use the Directional Information in Round Dances to Find Nearby Food Sources?" *Animal Behaviour* 83 (2012): 1319–24.

Haldane, John B. S., and Helen Spurway. "A Statistical Analysis of Communication in 'Apis Mellifera' and a Comparison with Communication in Other Animals." *Insectes Sociaux* 1 (1954): 247–83.

Harris, Randy Allen. *The Linguistics Wars*. New York: Oxford University Press, 1993.

Hasler, Arthur. Foreword to *Olfactory Imprinting and Homing in Salmon*, by Arthur Hasler and Allan T. Scholz, xi–xiii. Madison: University of Wisconsin Press, 1983.

———. Review of *Duftgelenkte Bienen im Dienste der Landwirtschaft und Imkerei*, by Karl von Frisch. *Science* 108, 2802 (1948): 290.

———. "This Is the Enemy." *Science* 102, 2652 (1945): 431.

———. "A War Time View of Some Biological Stations." *Biologist* 28 (1946): 81–93.

Hatzinger, Martin. "Anton Ritter von Frisch (1849–1917), Leben und Werk des ersten Präsidenten der DGU." *Der Urologe* 50 (2011): 719–21.

Hayes, Cathy. *The Ape in Our House*. New York: Harper, 1951.

Heim, Susanne. "Research for Autarky: The Contribution of Scientists to Nazi Rule in Germany." In *Ergebnisse, Vorabdrucke aus dem Forschungsprogramm Geschichte der Kaiser-Wilhelm-Gesellschaft im Nationalsozialismus*, edited by Carola Sachse, 4–30. Berlin: Max-Planck-Gesellschaft zur Förderung der Wissenschaften, 2001.

Heinroth, Katharina. *Mit Faltern begann es: Mein Leben mit Tieren in Breslau, München und Berlin*. Munich: Kindler, 1979.

Hentschel, Klaus, ed. *Physics and National Socialism: An Anthology of Primary Sources*. Translated by Ann M. Hentschel. Basel: Birkhäuser, 1996.

Herf, Jeffrey. *Reactionary Modernism: Technology, Culture, and Politics in Weimar and the Third Reich*. Cambridge: Cambridge University Press, 1984.

Hertwig, Oscar, and Richard Hertwig. *Die Actinien: Anatomisch und histologisch, mit besonderer Berücksichtigung des Nervenmuskelsystems*. Jena: G. Fischer, 1879.

Hess, Carl von. "Beiträge zur Frage nach einem Farbensinne bei Bienen." *Pflüger's Archiv für die gesammte Physiologie des Menschen und der Tiere* 170, 7–9 (1918): 337–66.

———. "Experimentelle Untersuchungen zur vegleichenden Physiologie des Gesichtssinnes." *Pflüger's Archiv für Physiologie* 142 (1910): 405–46.

———. "Der Gesichtssinn." In *Handbuch der vergleichenden Physiologie*, edited by Hans Winterstock, 555–840. Jena: Gustav Fischer Verlag, 1913.

———. "Neue Untersuchungen zur vergleichenden Physiologie des Gesichtssinnes." *Zoologische Jahrbücher* 33 (1913): 387–440.

———. "Untersuchungen über den Lichtsinn bei Fischen." *Archiv für Augenheilkunde*, 1909, 1–38.

———. "Untersuchungen zur Frage nach dem Vorkommen von Farbensinn bei Fischen." *Zoologische Jahrbücher* 31 (1912): 629–46.

Hobsbawm, Eric. *The Age of Extremes: Europe's Short Twentieth Century, 1914–1991*. New York: Vintage Books, 1996.

Hockett, Charles F. *A Course in Modern Linguistics*. New York: Macmillan, 1958.

Hofer, Veronika. "Rudolf Goldscheid, Paul Kammerer und die Biologen des Prater-Vivariums in der liberalen Volksbildung der Wiener Moderne." In *Wissenschaft, Politik und Öffentlichkeit: Von der Wiener Moderne bis zur Gegenwart*, edited by Mitchell G. Ash and Christian Stifter, 149–84. Vienna: WUV, 2002.

Hoffmann, Christoph, "Retenir et Reprendre: L'Écriture dans la Pratique de Recherche du Zoologue Karl von Frisch." *Genesis: Revue internationale de critique génétique* 36 (2013): 189–99.

Hopke, Christoph, ed. *Medizin und Verbrechen: Fesitschrift zum 60. Geburtstag von Walter Wuttke.* Ulm: Klemm & Oelschläger, 2001.

Huber, Francis [François]. *Huber's New Observations upon Bees: The Complete Volumes I & II.* Translated by C. P. Dadant. Nehawka, NE: X-Star, 2012.

Huber, Ursula. "Die Universität München—Ein Bericht über den Fortbestand nach 1945." In *Trümmerzeit in München: Kultur und Gesellschaft einer deutschen Großstadt im Aufbruch, 1945-1949,* edited by Friedrich Prinz, 156–60. Munich: C. H. Beck, 1984.

Jacobsen, Rowan. *Fruitless Fall: The Collapse of the Honey Bee and the Coming Agricultural Crisis.* New York: Bloomsbury, 2009.

Jaeger, Siegfried. "Vom erklärbaren, doch ungeklärten Abbruch einer Karriere— Die Terpsychologin und Sinnesphysiologin Mathilde Hertz (1891–1975)." In *Untersuchungen zur Geschichte der Psychologie und der Psychotechnik,* edited by H. Gundlach, 229–62. Munich: Profil Verlag, 1996.

Jansen, Sarah. *"Schädlinge" Geschichte eines wissenschaftlichen und politischen Konstrukts 1840–1920.* Frankfurt am Main: Campus, 2003.

Jardine, Lisa. *Ingenious Pursuits: Building the Scientific Revolution.* New York: Anchor Books, 1999.

Johnson, Dennis L. "Honey Bees: Do They Use the Direction Information Contained in Their Dance Maneuvers?" *Science* 155 (1967): 844–47.

Kak, Subhash. "The Honey Bee Dance Language Controversy." *Mankind Quarterly,* Summer 1991, 357–65.

Kalmár, János, and Alfred Stalzer, eds. *Das Jüdische Wien.* Vienna: Pichler, 2000.

Keller, Gottfried, Marie von Frisch-Exner, Adolf Exner, and Irmgard Smidt. *Aus Gottfried Kellers glücklicher Zeit: Der Dichter im Briefwechsel mit Marie und Adolf Exner.* Stäfa: T. Gut, 1981.

Kershaw, Ian. *Hitler, 1889–1936: Hubris.* New York: W. W. Norton, 1998.

Kevles, Daniel J. *In the Name of Eugenics: Genetics and the Uses of Human Heredity.* New York: Knopf, 1985.

Kirby, David. "Science Consultants, Fictional Films, and Scientific Practice." *Social Studies of Science* 33, 2 (2003): 231–68.

Klueting, Edeltraud. "Die gesetzliche Regelung der Nationalsozialistischen Reichsregierung für den Tierschutz, den Naturschutz und den Umweltschutz." In *Naturschutz und Nationalsozialismus,* edited by Joachim and Frank Ueköttter Radkau, 77–106. Frankfurt am Main: Campus, 2003.

Koestler, Arthur. *The Case of the Midwife Toad.* London: Hutchinson, 1971.

Kosek, Jake. "Ecologies of Empire: On the New Uses of the Honeybee." *Cultural Anthropology* 25, 4 (2010): 650–78.

———. *Homo Apians: A Critical Natural History of the Modern Honeybee.* Durham, NC: Duke University Press, forthcoming.

Krames, Lester, Patricia Pliner, and Thomas Alloway, eds. *Nonverbal Communication.* New York: Plenum Press, 1974.

Kressley-Mba, Regina A., and Siegfried Jaeger. "Rediscovering a Missing Link: The Sensory Physiologist and Comparative Psychologist Mathilde Hertz

(1891–1975)." *History of Psychology* 6, 4 (2003): 379–96.

Kreutzer, Ulrich. *Karl von Frisch (1886–1982): Eine Biografie.* Munich: August Dreesbach, 2010.

Krige, John. *American Hegemony and the Postwar Reconstruction of Science in Europe.* Cambridge, MA: MIT Press, 2006.

Large, David Clay. *Where Ghosts Walked: Munich's Road to the Third Reich.* New York: W. W. Norton, 1997.

Lavédrine, Bertrand, and Jean-Paul Gandolfo. *The Lumière Autochrome: History, Technology, and Preservation.* Translated by John McElhone. Los Angeles: Getty Conservation Institute, 2013.

Lemov, Rebecca. *World as Laboratory: Experiments with Mice, Mazes, and Men.* New York: Hill and Wang, 2005.

Likens, Gene E. "Arthur David Hasler, 1908–2001: A Biolgraphical Memoir." *Biographical Memoirs* 82 (2002): 1–14.

Lindauer, Martin. *Communication among Social Bees.* 1961. Cambridge, MA: Harvard University Press, 1971.

———. "Schwarmbienen auf Wohnungssuche." *Zeitschrift für vergleichende Physiologie* 37, 4 (1955): 266–324.

Lipphardt, Veronika. *Biologie der Juden: Jüdische Wissenschaftler über "Rasse" und Vererbung 1900–1935.* Göttingen: Vandenhoeck & Ruprecht, 2008.

Litten, Freddy. *Der Rücktritt Richard Willstätters 1924/25 und seine Hintergründe: Ein Münchner Universitätsskandal?* Munich: Münchener Universitätsschriften, 1999.

———. "Die 'Verdienste' eines Rektors im Dritten Reich." *N.T.M* 11, 1 (2003): 34–46.

Lotz, John. "'Karl Von Frisch: Bees, Their Vision, Chemical Senses, and Language,' Book Review." *Word* 7 (1951): 66–67.

———. "Speech and Language." *Journal of the Acoustical Society of America* 22 (1950): 712–17.

Lubbock, John. *Ants, Bees, and Wasps: A Record of Observations on the Habits of the Social Hymenoptera.* 7th ed. London: Kegan, Paul, Trench, 1885. First published 1881.

Maclaurin, Colin. "On the Bases of the Cells Wherein the Bees Deposite Their Honey." *Philosophical Transactions of the Royal Society of London* 42 (1742): 565–71.

MacLuhan, Marshall. *The Medium Is the Message.* New York: Random House, 1967.

Mandler, George. "Origins of the Cognitive Revolution." *Journal of the History of the Behavioral Sciences* 38 (2002): 339–53.

Manning, Aubrey. *An Introduction to Animal Behaviour.* 3rd ed. Reading, MA: Addison-Wesley, 1979. First edition 1967.

Mazower, Mark. *Dark Continent: Europe's Twentieth Century.* New York: Vintage Books, 1998.

Mazzolini, Renato G. "Adam Gottlob Schirach's Experiments on Bees." In *The Light of Nature,* edited by J. D. North and J. J. Roche, 67–82. Amsterdam: Marinus Nijhoff Publishers, 1985.

Melzer, Jörg. *Vollwerternährung. Diätetik, Naturheilkunde, Nationalsozialismus, sozialer Anspruch.* Stuttgart: Steiner, 2003.

Merrick, Jeffrey. "Royal Bees: The Gender Politics of the Beehive in Early Modern Europe." *Studies in Eighteenth-Century Culture* 18 (1988): 7–37.

Meyer, Beate. *"Jüdische Mischlinge": Rassenpolitik und Verfolgungserfahrung 1933–1945,* Studien zur jüdischen Geschichte Bd. 6. Hamburg: Dölling und Galitz, 1999.

Montgomery, Georgina Mary. "Place, Practice and Primatology: Clarence Ray Carpenter, Primate Communication and the Development of Field Methodology, 1931–1945." *Journal of the History of Biology* 38, 3 (2005): 495–533.

Müller-Hill, Benno. *Murderous Science: Elimination by Scientific Selection of Jews, Gypsies, and Others, Germany 1933–1945.* Translated by George Fraser. Oxford: Oxford University Press, 1988.

Müller-Wille, Staffan, and Hans-Jörg Rehinberger. *Heredity Produced: At the Crossroads of Biology, Politics, and Culture, 1500–1870.* Cambridge, MA: MIT Press, 2007.

Munz, Tania. "The Bee Battles: Karl Von Frisch, Adrian Wenner and the Honey Bee Dance Language Controversy." *Journal of the History of Biology* 38 (2005): 535–70.

———. "Die Ethologie des wissenschaflichen Cineasten." Translated by C. Brinckmann and S. Lowry. *montage/av* 14, 4 (2005): 52–68.

Nagel, Anne. "'Er ist der Schrecken überhaupt der Hochschule'—Der Nationalsozialistische Deutsche Dozentenbund in der Wissenschaftspolitik des Dritten Reichs." In *Universitäten und Studenten im Dritten Reich, Bejahung, Anpassung, Widerstand. XIX. Königswinterer Tagung vom 17.–19. Februar 2006,* edited by Joachim Scholtyseck and Christoph Studt, 115–32. Berlin: LIT, 2008.

Nyhart, Lynn K. *Biology Takes Form: Animal Morphology and the German Universities, 1800–1900.* Chicago: University of Chicago Press, 1995.

Paul, Diane B. *Controlling Human Heredity: 1865 to the Present.* Control of Nature. Atlantic Highlands, NJ: Humanities Press, 1995.

Pauly, Philip J. *Controlling Life: Jacques Loeb and the Engineering Ideal in Biology.* New York: Oxford University Press, 1987.

Paxton, Robert O., and Julie Hessler. *Europe in the Twentieth Century.* Boston: Wadsworth, Cengage Learning, 2012.

Prete, Frederick R. "Can Females Rule the Hive? The Controversy over Honey Bee Gender Roles in Beekeeping Texts of the Sixteenth–Eighteenth Centuries." *Journal of the History of Biology* 24 (1991): 113–44.

Prodger, Phillip. *Time Stands Still: Muybridge and the Instantaneous Photography Movement.* New York: Oxford University Press, 2003.

Przibram, Hans. "Die Biologische Versuchsanstalt in Wien: Ausgestaltung und Tätigkeit während des zweiten Quinquenniums ihres Bestandes (1908–1912), Bericht der zoologischen, botanischen und physikalisch-chemischen Abteilung." *Zeitschrift für biologische Technik und Methodik* 3 (1913): 163–245.

———. "Die Biologische Versuchsanstalt in Wien: Zweck, Einrichtung und Tätigkeit während der ersten fünf Jahre ihres Bestandes (1902–1907), Bericht der zoologischen, botanischen und physikalisch-chemischen Abteilung" [in

4 parts]. *Zeitschrift für biologische Technik und Methodik* 1 (1908): 234–64, 329–62, 409–33; and (1909): 1–34.

Radick, Gregory. "Morgan's Canon, Garner's Phonograph, and the Evolutionary Origins of Language and Reason." *British Journal of the History of Medicine* 33, 1 (2000): 3–23.

———. *The Simian Tongue: The Long Debate about Animal Language.* Chicago: University of Chicago Press, 2008.

Raffles, Hugh. *The Illustrated Insectopedia.* New York: Pantheon Books, 2010.

Reid, Thomas. *The Works of Thomas Reid.* 7th ed. Edinburgh: Maclachlan and Stewart, 1882.

Reinhardt, Klaus. "The Entomological Institute of the Waffen-SS." *Endeavour* 37, 4 (2013): 220–27.

Reiter, Wolfgang L. "Zerstört und Vergessen: Die Biologische Versuchsanstalt und ihre Wissenschaftler/innen." *Österreichische Zeitschrift für Geschichtswissenschaften* 4 (1999): 585–614.

Remy, Steven P. *The Heidelberg Myth: The Nazification and Denazification of a German University.* Cambridge, MA: Harvard University Press, 2003.

Rentschler, Eric. *The Ministry of Illusion: Nazi Cinema and Its Afterlife.* Cambridge, MA: Harvard University Press, 1996.

Révész, Géza. "Der Kampf um die sogenannte Tiersprache." *Psychologische Rundschau* 4 (1953): 81–83.

Richards, Robert J. *Darwin and the Emergence of Evolutionary Theories of Mind and Behavior.* Chicago: University of Chicago Press, 1987.

Romanes, George John. *Animal Intelligence.* London: Kegan Paul, Trench, 1882.

Rösch, Gustav. "Über die Bautätigkeit im Bienenvolk und das Alter der Baubienen." *Zeitschrift für vergleichende Physiologie* 6 (1927): 264–98.

———. "Untersuchungen über die Arbeitsteilung im Bienenstaat I." *Zeitschrift für vergleichende Physiologie* 2 (1925): 571–631.

———. "Untersuchungen über die Arbeitsteilung im Bienenstaat II." *Zeitschrift für vergleichende Physiologie* 12 (1930): 1–71.

Ruestow, Edward G. *The Microscope in the Dutch Republic: The Shaping of Discovery.* Cambridge: Cambridge University Press, 1996.

Russell, Edmund P., III. "'Speaking of Annihilation': Mobilizing for War against Human and Insect Enemies." *Journal of American History* 82, 4 (1996): 1505–29.

Sachse, Carola. "What Research, to What End? The Rockefeller Foudnation and the Max Planck Gesellschaft in the Early Cold War." *Central European History* 42 (2009): 97–141.

Sachse, Carola, and Mark Walker, eds. "Politics and Science in Wartime: Comparative International Perspectives on the Kaiser Wilhelm Institute." *Osiris* 20 (2005).

Santschi, Felix. "Observations et remarques critiques sur le mecanisme de l'orientation chez les fourmis." *Revue Suisse de Zoologie* 19 (1911): 29–70.

Saraiva, Tiago, and M. Norton Wise, "Autarky/Autarchy: Genetics, Food Production, and the Building of Fascism." *Historical Studies in the Natural Sciences* 40 (November 2010): 419–28.

Sax, Boria. *Animals in the Third Reich: Pets, Scapegoats, and the Holocaust.* New York: Continuum, 2000.

Schorske, Carl E. *Fin-de-Siècle Vienna: Politics and Culture*. New York: Vintage Books, 1981.

Sebeok, Thomas A., ed. *Animal Communication: Techniques of Study and Results of Research*. Bloomington: Indiana University Press, 1968.

———. "Communication among Social Bees; Porpoises and Sonar; Man and Dolphin." *Language* 39 (1963): 448–66.

Sebeok, Thomas A., and Alexandra Ramsay, eds. *Approaches to Animal Communication*. The Hague: Mouton, 1969.

Seeley, Thomas D. *Honeybee Democracy*. Princeton, NJ: Princeton University Press, 2010.

Seeley, Thomas D., S. Kühnholz, and R. H. Seeley, "An Early Chapter in Behavioral Physiology and Sociobiology: The Science of Martin Lindauer." *Journal of Comparative Physiology* 188 (2002): 439–53.

Sewig, Claudia. *Der Mann, der die Tiere liebte: Bernhard Grzimek*. Bergisch Gladbach: Lübbe, 2009.

Seyfarth, Ernst-August, and Henryk Perzchala. "Sonderaktion Krakau 1939: Die Verfolgung von polnischen Biowissenschaftlern und Hilfe durch Karl von Frisch." *Biologie in unserer Zeit* 22, 4 (1992): 218–25.

Shapin, Steven. "Pump and Circumstance: Robert Boyle's Literary Technology." *Social Studies of Science* 14, 4 (1984): 481–552.

———. *The Scientific Revolution*. Chicago: University of Chicago Press, 1998.

Shapin, Steven, and Simon Schaffer. *Leviathan and the Air-Pump: Hobbes, Boyle, and the Experimental Life*. Princeton, NJ: Princeton University Press, 1985.

Sinclair, Upton. *The Jungle*. Clayton: Prestwich House, 2005.

Sivin, Nathan. "Why Did the Scientific Revolution Not Take Place in China—or Didn't It?" *Chinese Science* 5, 1 (1982): 45–66.

Sleigh, Charlotte. *Six Legs Better: A Cultural History of Myrmecology*. Baltimore, MD: Johns Hopkins University Press, 2007.

Smith, David. "'Cruelty of the Worst Kind': Religious Slaughter, Xenophobia, and the German Greens." *Central European History* 40 (2007): 89–115.

Smith, Roger. *Norton History of the Human Sciences*. New York: W. W. Norton, 1997.

Solnit, Rebecca. *River of Shadows: Eadweard Muybridge and the Technological Wild West*. New York: Viking, 2003.

Sprengel, Christian Conrad. *Das entdeckte Geheimnis der Natur im Bau und in der Befruchtung der Blumen*. Berlin: Vieweg, 1793.

Stellwaag, Friedrich. "Über die Beziehung des Lebens zum Licht." *Muenchener Medizinische Wochenschrift* 62, 48 (1915): 1642–43.

Swammerdam, Jan. *The Book of Nature or, the History of Insects: Reduced to Distinct Classes, Confirmed by Particular Instances, . . . and Illustrated with Copper-Plates. . . . By John Swammerdam, M.D. With the Life of the Author, by Herman Boerhaave, M.D. Translated . . . by Thomas Flloyd. Revised and improved . . . by John Hill, M.D.* London: S. G. Seyffert, 1758.

Taschwer, Klaus. "From the Aquarium to the Zoo to the Lab." In *A Laboratory in the Prater: Experiment and Theory in the Biologische Versuchsanstalt in Vienna*, edited by Sabine Brauckmann and Gerd Müller. Cambridge, MA: MIT Press, forthcoming.

———. *Hochburg des Antisemitismus: Der Niedergang der Universität Wien im 20. Jahrhundert*. Vienna: Czernin Verlag, 2015.

Tautz, Jürgen. *Die Erforschung der Bienenwelt: Neue Daten—neues Wissen*. 2nd ed. Stuttgart: Klett MINT Verlag, 2015.

Tent, James F. *Mission on the Rhine: "Reeducation" and Denazification in American-Occupied Germany*. Chicago: University of Chicago Press, 1982.

Thomas, Marion. "Are Animals Just Noisy Machines? Louis Boutan and the Co-Invention of Animal and Child Psychology in the French Third Republic." *Journal of the History of Biology* 38, 3 (2005): 425–60.

Thorpe, William H. "Orientation and Methods of Communication of the Honey Bee and Its Sensitivity to the Polarization of the Light." *Nature* 164, 4157 (1949): 11–14.

———. *The Origins and Rise of Ethology: The Science of the Natural Behaviour of Animals*. London: Heinemann Educational Books, 1979.

Turner, R. Steven. *In the Eye's Mind: Vision and the Helmholtz-Hering Controversy*. Princeton, NJ: Princeton University Press, 1994.

Veldink, Connie J. "The Honey-Bee Language Controversy." *Interdisciplinary Science Reviews* 14 (1989): 166–75.

Vicedo, Marga. *The Nature and Nurture of Love: From Imprinting in Ducks to Attachment in Infants*. Chicago: University of Chicago Press, 2012.

Vogt, Annette. *Wissenschaftlerinnen in Kaiser-Wilhelm-Instituten, A–Z: Veröffentlichungen aus dem Archiv zur Geschichte der Max Planck Gesellschaft*. Berlin: Max Planck Gesellschaft, 2008.

von Frisch, Karl. *See* Frisch, Karl von.

von Hess, Carl. *See* Hess, Carl von.

Walker, Mark. *Nazi Science: Myth, Truth, and the German Atomic Bomb*. New York: Plenum Press, 1995.

Watson, John Broadus. *Behaviorism*. 1924. New Brunswick, NJ: Transaction Publishers, 1997.

Weindling, Paul. *Health, Race, and German Politics between National Unification and Nazism, 1870–1945*. Cambridge History of Medicine. Cambridge: Cambridge University Press, 1989.

Weinreb, Alice. *Modern Hungers: A Political Economy of Food in 20th-Century Germany*. Oxford: Oxford University Press, forthcoming.

Wells, Patrick, and Adrian Wenner, "Do Honey Bees Have a Language?" *Nature* 241 (1973): 171–74.

Wenner, Adrian. "Honey Bees." In *Animal Communication: Techniques of Study and Methods of Research*, edited by Thomas A. Sebeok, 217–43. Bloomington: Indiana University Press, 1968.

———. "Honey Bees: Do They Use the Distance Information Contained in Their Dance Maneuver?" *Science* 155 (1967): 847–49.

———. "Information Transfer in Honey Bees: A Population Approach." In *Nonverbal Communication*, edited by Lester Krames, Patricia Pliner, and Thomas Alloway, 133–69. New York: Plenum Press, 1974.

Wenner, Adrian, and Patrick Wells. *Anatomy of a Controversy*. New York: Columbia University Press, 1990.

Wenner, Adrian, Patrick O. Wells, and Dennis L. Johnson, "An Analysis of the

Waggle Dance and Recruitment in Honey Bees." *Physiological Zoölogy* 40, 4 (1967): 317–44.

———. "Honeybee Recruitment to Food Sources: Olfaction or Language?" *Science* 164 (1969): 84–86.

Wheeler, William Morton. *Ants: Their Structure, Development and Behavior.* New York: Columbia University Press, 1910.

Wilson, Edward O. "Animal Communication." *Scientific American* 227 (1972): 52–60.

———. *The Insect Societies.* Cambridge, MA: Belknap Press of Harvard University Press, 1971.

———. "Letters." *Scientific American* 227 (1972): 52–60.

Wolf, Ernst. "Über das Heimkehrvermögen der Bienen, II." *Zeitschrift für vergleichende Physiologie* 6, 2 (1927): 221–54.

Zerner, Charles. "Stealth Nature: Biomimesis and the Weaponization of Life." In *In the Name of Humanity: The Government of Threat and Care*, edited by Ilana Feldman and Miriam Ticktin, 290–324. Durham, NC: Duke University Press, 2010.

Ziemke, Earl. *The US Army and the Occupation of Germany, 1944–1946.* Washington, DC: Center of Military History, United States Army, 1975.

## Films

von Frisch, Karl. *Entwicklung der Honigbiene und des Bienenvolkes* [Development of the honeybee and of a colony]. 8.5 minutes. Germany: IWF (Formerly: Institut für Film und Bild in Wissenschaft und Unterricht, Göttingen), 1942–1944.

———. *Farbensinn der Bienen* [Sense of color in bees]. 7.5 minutes. Germany: IWF Wissen und Medien (Formerly: RfdU, Berlin), 1926.

———. *Geruchsinn der Bienen* [Sense of smell in bees]. 8.5 minutes. Germany: IWF Wissen und Medien (Formerly: RWU, Berlin), 1927.

———.*Geschmackssinn bei Fischen* [Sense of taste in fish]. 4.5 minutes. Germany: IWF Wissen und Medien (Formerly: RWU, Berlin), 1927.

———. *Geschmackssinn der Bienen* [Sense of taste in bees]. 3.5 minutes. Germany: IWF Wissen und Medien (Formerly: RWU, Berlin), 1927.

———. *Hörvermögen der Elritze* (Phoxinus laevis) [The sense of hearing in the minnow]. 6.5 minutes. Germany: IWF Wissen und Medien, 1929.

———. *Pollen- und Nektarsammeln der Honigbiene* [Collection of pollen and nectar in the honeybee]. Germany: IWF Wissen und Medien (Formerly: Institut für Film und Bild in Wissenschaft und Unterricht, Göttingen), 1942–1944.

———. *Sprache der Bienen* [Language of bees]. 8 minutes. Germany: IWF Wissen und Medien, 1926.

Harlan, Veit. *Jud Süß*. Germany: Terra-Filmkunst GmbH, 1940.

Hippler, Fritz. *Der ewige Jude: Dokumentarfilm über das Weltjudentum*. Germany: Deutsche Filmherstellungs- und -Verwertungs-GmbH, Berlin, 1940.

Liebeneiner, Wolfgang. *Ich klage an*. Germany: Tobis-Filmkunst GmbH, Berlin, 1941.

## PhD Dissertations

Henkel, Christian. "Unterscheiden die Bienen Tänze?" PhD diss., University of Bonn, 1938.

Veldink, Connie J. "Paradigm Challenges in Modern Science: The Bee Language Controvery." PhD diss., University of California, Santa Barbara, 1976.

Wiecki, Stefan Wolfgang. "Professors in Purgatory: The Denazification of Munich University, 1945–1955." PhD. diss., Brandeis University, 2009.

# Index